Fatal Evidence

One day a shepherd was crossing the bridge when he saw a little bone beneath him in the sand. It was so pure and snow-white that he wanted it to make a mouthpiece from, so he climbed down and picked it up. Afterward he made a mouthpiece from it for his horn, and when he put it to his lips to play, the little bone began to sing by itself:

> *Oh, dear shepherd*
> *You are blowing on my bone.*
> *My brothers struck me dead,*
> *And buried me beneath the bridge.*

From 'The Singing Bone', collected by The Brothers Grimm.
Translated by Professor D.L. Ashliman

Fatal Evidence

Professor Alfred Swaine Taylor & the Dawn of Forensic Science

Helen Barrell

PEN & SWORD
HISTORY

First published in Great Britain in 2017 by
Pen & Sword History
an imprint of
Pen & Sword Books Ltd
47 Church Street
Barnsley
South Yorkshire
S70 2AS

ISBN 978 1 47388 341 3

Typeset in Ehrhardt by
Mac Style Ltd, Bridlington, East Yorkshire

Printed and bound in Malta by Gutenberg Press Ltd.

Pen & Sword Books Ltd incorporates the imprints of Pen & Sword
Archaeology, Atlas, Aviation, Battleground, Discovery, Family History,
History, Maritime, Military, Naval, Politics, Railways, Select, Transport,
True Crime, Fiction, Frontline Books, Leo Cooper, Praetorian Press,
Seaforth Publishing and Wharncliffe.

For a complete list of Pen & Sword titles please contact
PEN & SWORD BOOKS LIMITED
47 Church Street, Barnsley, South Yorkshire, S70 2AS, England
E-mail: enquiries@pen-and-sword.co.uk
Website: www.pen-and-sword.co.uk

Contents

Note on Text		vi
Introduction		vii
Chapter 1	Go Thy Way, Passenger	1
Chapter 2	More of Impulse than Discretion	12
Chapter 3	Fearful and Wonderful	19
Chapter 4	The Light of an English Sun	26
Chapter 5	One of the Most Eminent Men	36
Chapter 6	My Heart is as Hard as a Stone	48
Chapter 7	The Means of our Preservation	58
Chapter 8	The Only Friend I had in the World	69
Chapter 9	The Formidable Scourge	79
Chapter 10	His Very High Position	91
Chapter 11	Romantic, Mysterious, and Singular	103
Chapter 12	Enter Not into the Path of the Wicked	113
Chapter 13	Truth Will Always Go the Farthest	143
Chapter 14	Grieved Beyond all Endurance	151
Chapter 15	You are the Villain	166
Chapter 16	Blood Enough	179
Chapter 17	The Eminent Opinion of Professor Taylor	190
Timeline		204
Acknowledgements		206
Further Reading		207
Selected Bibliography		208
Notes		213
Index		225

Note on Text

Medical jurisprudence

Taylor called himself a 'medical jurist' practising 'medical jurisprudence' – the intersection of medicine and the law. He sometimes described himself as a toxicologist. Had he lived today, he would have been called a pathologist, or a forensic science technician.

The Old Bailey

The Central Criminal Court stands on a London street called the Old Bailey; this is the name by which the court is commonly known. In newspaper and trial reports, the name can switch about from the Central Criminal Court to the Old Bailey. To avoid confusion, the Old Bailey is used throughout.

Inquests

In England and Wales, sudden, violent, or unnatural deaths are investigated by a coroner. In Taylor's day, a jury would hear the evidence and come to a verdict. On finding wilful murder against a particular individual, the case would go to trial if a grand jury decided that it was strong enough. Even if the inquest jury's verdict was wilful murder, the case could still be thrown out and the accused could go free, or be tried on a lesser charge. If the jury found that the death was the result of natural causes but the coroner strongly suspected otherwise, further investigations would be carried out by police and magistrates.

The Crown

Under the legal system in England and Wales, the prosecution is said to be the Crown (hence the trial of William Palmer is R v Palmer; the R standing for 'Regina', Latin for 'Queen'). Prosecution witnesses are therefore 'witnesses for the Crown'. To avoid confusion for international readers, 'prosecution' is used in this book instead of 'the Crown'. Taylor often used 'prosecution' instead of 'Crown' in his writing.

Introduction

A body has been found. There are severe head injuries, as if the victim has been trampled to death. It's manslaughter, possibly murder. There's a suspect, and their hobnail boots are covered in something red. Could it be blood? The boots, and snippets of the victim's hair, are sent to a laboratory for analysis. It *is* blood on the boots, and there are fibres in the mud around the hobnails, which, under a microscope, match the victim's hair. There are strands of red wool in the coagulated blood, and further information says that the victim wore a red scarf. The analysis has shown that the victim was probably killed by whoever wore the boots, and the suspect is arrested.

This did not happen last week, neither is it a scene from a crime drama on television. It took place in 1863, and the analysis was performed by Dr Alfred Swaine Taylor, professor of medical jurisprudence, toxicology and chemistry at Guy's Hospital in London.

Codes and laws applied to medicine and the examination of violent death go back a long way in human history, as far as Ancient Babylon. It was in Renaissance Europe that legal codes around medicine and violent death were further developed, creating the basis for what we recognise as forensic science today. Britain lagged behind until the early nineteenth century, when advances in science and technology led to the merging of medicine and chemistry and the development of a new science: forensic medicine. Alfred Swaine Taylor, skilled as a chemist and a physiologist, was one of the first, and youngest, lecturers of medical jurisprudence in England.

Taylor is remembered today as a toxicologist, involved in the sensational trials of medical men Palmer and Smethurst. They were notoriously difficult cases. Palmer was thought to have poisoned with strychnine, and scientific opinion differed angrily as to how it could be traced. In Smethurst's case, a test for arsenic that was thought to be infallible was shown to be flawed. But there is far more to Taylor than that.

Taylor worked on hundreds of cases from the 1830s to the 1870s, rising to prominence in England as a leading expert. It was, perhaps, his prolific writing output – his *cacoethes scribendi*, as one of his rivals put it – that helped him to assume his respected position. His opinion was sought by the Home Office on difficult cases, and local newspapers proudly announced if a coroner had summoned Taylor's aid. He became a household name.

No diaries of Taylor's have survived, and neither have any bundles of personal letters, although the occasional piece of correspondence turns up in archives. He was so busy writing books and articles that he had no time to write his memoirs. But the curious can piece together his professional life through newspaper reports of the inquests and trials that he appeared at, and the inside information that sometimes slips through in his books and articles. Glimpses of his personality flash through in his writing: his scathing sarcasm aimed at his professional peers who had the temerity to cross him, his fondness for wordplay, and his rage at the stubbornness of governments who prioritised commerce over public safety. Genealogical records allow us to occasionally press our noses up against the window of his home. For the most part it was a haven, even if tragedy visited his wife's family, and the occasional policeman and surgeon knocked at his door with a grim burden for his analysis.

This is both Taylor's biography and the story of forensic science's development in nineteenth-century England; the two are entwined. There are stomachs in jars, a skeleton in a carpet bag, doctors gone bad, bloodstains on floorboards, and an explosion that nearly destroyed two towns. This is the true tale of Alfred Swaine Taylor and his fatal evidence.

Chapter 1

Go Thy Way, Passenger
1806–31

Take notice, roguelings

Alfred Swaine Taylor was born on 11 December 1806, in Northfleet, Kent, the first child of Thomas and Susannah Taylor. The small town on the chalky banks of the river Thames is about 25 miles east from the centre of London, and was known in the early nineteenth century for its watercress and flint. Taylor's maternal grandfather, Charles Badger, had been a wealthy local flint knapper, supplying the essential component for flintlock pistols. Taylor's father was a captain in the East India Company, and perhaps moved to Northfleet from his native Norfolk when, in 1804, the company leased twelve moorings in the town for its ships.

Taylor's middle name, Swaine, presumably harks back to an ancestor on his father's side as he had several relatives with the same middle name. Just after Taylor's second birthday, he was joined by a brother, Silas Badger Taylor. There were to be only two Taylor children; their mother died aged 37, in December 1815, three days before Taylor's ninth birthday. Her headstone can still be found in the churchyard at Northfleet. It reads: 'She was worthy of example as a Wife a Mother and a Friend. Go thy way passenger and imitate Her whom you will some day follow.'

By 1818, Taylor's father had become a merchant, half of Oxley and Taylor, who were based at the chalk works in Northfleet. They advertised themselves as 'exporters and dealers in all sorts of flints, rough or manufactured, chalk etc, to India, and all parts of England'. The company would later expand to include an office off Lombard Street in London, and Silas would grow up to continue their father's business. By the 1830s, Oxley and Taylor would sell guns, and were travel agents, selling berths on ships bound for New York.

Taylor was a studious boy, not considered strong enough for the rough and tumble life of a public school. Soon after his mother's death, he was sent to Albemarle House, the school of clergyman Dr Joseph Benson. Located in Hounslow, on a major stagecoach route to the west of London, it was a journey of about 40 miles from Northfleet. Pupils from across the country could reach the school with ease, and its semi-rural location was healthier than the city or town. Albemarle House, 'a school of the first respectability', had been established at some point during the mid-eighteenth century; it was a large, imposing Georgian edifice.

The building has since disappeared, as have the gibbets that once stood very near the school by the junction of two major roads. The bodies of executed highwaymen swung there as a warning to anyone who thought to ply their trade on Hounslow Heath. They made a terrifying sight: 'The chains rattled; the iron plates barely held the gibbet together; the rags of the highwaymen displayed their horrible skeletons within.' The gibbets could be seen with ease from the school until their removal in about 1806. In the late eighteenth century, a jocular wag wrote a six-line poem, 'Addressed to Two Young Gentlemen at the Hounslow Academy':

> *Take notice, roguelings, I prohibit*
> *Your walking underneath yon Gibbet;*
> *Have you not heard, my little ones,*
> *Of* Raw Head and Bloody Bones?
> *How do you know but that there fellow,*
> *May step down quick, and up you swallow?*

Although the gibbets had gone by the time Taylor attended Albemarle, their legend must have lingered on amongst the pupils at the school.

An aquatint of Albemarle House, from 1804, shows a military parade in action outside, even though it doesn't appear to have been a military academy. [Plate 2] The soldiers might have been pupils, drilled as volunteers for the war against Napoleon, or were professionals from Hounslow Camp, using space outside the school to practise. In the late 1830s, an advert for the school offered 'high Classical and Mathematical Reading, preparatory to the Universities or the Public Schools'. French was taught by a resident Parisian and instruction was given in 'English Literature, Commercial Science, and every subject of Education essential to the improvement of the Mind, and the refinement of the Understanding'. All of which would stand Taylor in good stead.

'Physiologist Taylor'

By 1822, Taylor's schooldays were over: in June that year, not yet 16, he was apprenticed to a surgeon. All manner of occupations could be entered by apprenticeship; Taylor's brother would be apprenticed to a London waterman.

There were various ways to enter medicine at this period, whether one wished to be a physician, apothecary or surgeon. Unless they attended the medical schools attached to Scottish universities, British physicians trained at universities in Continental Europe, briefly studying at Oxford or Cambridge to have the degree of MD conferred on them; their knowledge came mainly from books. As surgeons and apothecaries were apprenticed, they had more practical experience with the sick. There was rivalry between the three professions in London, the Guilds and Worshipful Companies protecting the interests of their members, sometimes at the expense of the sick. But at least it meant that medical professionals were regulated to an extent, and only the best candidates were selected as apprentices.

Taylor was apprenticed to Dr Donald McRae, a surgeon in Lenham, Kent, about 25 miles south-east of Northfleet. It was the birthplace of his maternal grandmother and it is possible that Taylor still had relatives there. McRae was in his thirties, originally from Inverness in Scotland. How proficient he was as a surgeon is unknown; by 1841 he had given up medicine for banking.

The terms of Taylor's apprenticeship stated that after a year of provincial medicine with McRae, he would become a pupil at a London hospital. In October 1823, Taylor entered the United Hospitals of Guy's and St Thomas's in Southwark, south London. Sometimes, provincial apprentice surgeons served part of their training in a hospital, usually spending six months to a year there. They gained invaluable experience, witnessing cases in numbers that wouldn't be seen in a rural practice, and they could perform dissections and observe operations. The studious Taylor would stay at Guy's for the rest of his career.

St Thomas's was an ancient hospital, founded by mediaeval monks for the treatment of the poor, on the Thames's south bank near London Bridge. In the early eighteenth century, one of St Thomas's wealthy governors, Thomas Guy, was urged by a physician friend to build a new hospital. Guy was aware that many of the patients were not well enough to work once they were deemed cured and had been discharged. So he established his hospital as a place for 'incurables' – a hospital for convalescence and rehabilitation. Guy's opened its doors in 1726, a year after its benefactor's death, occupying new buildings beside St Thomas's.

A medical school of sorts was run at Guy's from its inception. Its physicians, surgeons and apothecaries took on apprentices and pupils, and it had a surgical theatre from 1739 so that students could observe operations, even though few surgeries were performed. The hospital staff were voluntary, and were each paid £40 a year. The sum was stipulated at the foundation of Guy's and was only increased at the creation of the NHS. It was in the interest of staff to take on pupils and apprentices in order to increase their income.

A lecture theatre was an important lure for students, and one was built at Guy's in 1770. A fee was paid to be enrolled as a pupil, and they followed medical practitioners on their rounds. They paid extra for classes, lectures and dissection. Lectures in anatomy and surgery were most in demand, and were taught at St Thomas's. Guy's provided lectures in medicine, chemistry, botany, physiology and natural physiology.

When Taylor arrived at the United Hospitals in 1823, he took great interest in his chemistry lessons, taught by William Allen and Arthur Aikin. Allen was a Fellow of the Royal Society, highly respected in his time as a man of science, but he was not a medical man. Neither was Aikin, whose focus was on geology and coalfields, but both were amongst the leading chemists of their time. It has been said that Taylor was the favourite pupil of celebrated surgeon Sir Astley Cooper. Cooper was so famous that after he died, a commemorative statue was placed in St Paul's Cathedral, albeit with the wrong year of death on its plinth.

Taylor spent the summer of 1825 in Paris; cadavers for dissection were more easily acquired on the Continent. His trip paid off, because when he returned to London, he was awarded St Thomas's anatomy prize. He became known for his aptitude in physiology, earning himself the nickname 'Physiologist Taylor'. That same year, Guy's Medical School was founded, so an anatomical theatre, a dissecting room and a museum were built.

In 1826, Taylor's life was changed when he read *Elements of Medical Jurisprudence*, by American physician Theodric Romeyn Beck. Taylor later wrote, 'The subject, from the lucid manner in which it was treated by the author, fixed my attention, and induced me for the time to put aside anatomy and physiology for the sake of this new branch of medical science.' Reading Beck's book was the deciding moment in Taylor's career, when he chose medical jurisprudence as his 'special object for study and practice'.

In 1828, aged 21, he became a licentiate of the Society of Apothecaries, and he took off on a tour of medical schools in Europe. He returned to Paris, attending lectures by eminent men of the time, such as pathologist and haematologist Gabriel Andral; surgeons Guillaume Dupuytren and Alexis Boyer; physician and embryologist Étienne Serres; chemist and physicist Joseph Louis Gay-Lussac; and mineralogist and geologist Alexandre Brongniart. Considering Taylor's later specialisms, perhaps the most important lectures of all were those of toxicologist Joseph Bonaventura Orfila. He had published his two-volume work *Traité des poisons* from 1814 to 1815. His book explained the symptoms of poisoning, antidotes, and methods to detect poisons, as well as how to combine the post-mortem with analytical chemistry, thus merging all of Taylor's interests into one discipline.

A hollow crust spread over an abyss

Travel at this time was by ship or horse, but perhaps for reasons of economy, Taylor made much of his European journey on foot. Despite improvements in Continental travel by the late 1820s, such as better roads and fewer Italian *banditti*, Taylor's progress was not only slow but fraught with danger. His ship from France to Naples was racked by storm, and he was chased off Elba by pirates. He was arrested: once for having dangerous books, and then for espionage after he sketched some fortifications in northern Italy. He was only freed when most of his artwork was destroyed, and he was sent under escort to the border. Somehow, Taylor's sketch of Greek ruins in southern Italy survived; its careful intricacy shows his keen eye for detail.

He made his way round the medical schools and hospitals of Naples, Rome, Florence, Bologna, Milan, Heidelberg, Leiden, Amsterdam and Brussels. On his way, he travelled through the Auvergne, and hiked up the volcanic Puy de Dôme in central France. While in Naples, he wrote two ophthalmological articles in Italian for the *Giornale Medico Napolitano*; '*Sull'Inversione degli oggetti nel fondo del Occhio*' ('On inverting objects at the back of the eye'), and '*Sul conformarsi dell' occhio alla distanza degli oggetti*' ('On adapting the eye to the distance of objects').

He headed to the Solfatara, near Naples, an extinct volcano where sulphur and alum were procured, and looked around the refining factory. On hearing rumbles deep underground, he was told by guides that 'the Solfatara is even now but a hollow crust spread over an abyss,' but Taylor had heard the same rumbling at the Puy de Dôme and in the Alps. He assigned it to crumbling in certain parts of the crater; the way the noise travelled through the hillside made it sound more terrifying than it was.

Near the Solfatara was the Grotta del Cane, a small cave in the side of a hill, where an invisible, suffocating gas lay on its sloping floor in a lake, 'to about the height of the knee'. The gas ran up through the rocks from an extinct volcano, and was generally thought to be sulphurous. Taylor, however, had read up on it, and theorised that it was carbonic acid. Chemistry was not easy to study in Italy at the time, and with difficulty he had managed to get hold of limewater and phosphorus for his experiments. He believed that the results, including a torch extinguishing at the surface of the gas lake, proved his theory to be correct. Later experiments by others showed that it was carbon dioxide, but at least Taylor had demonstrated that the gas wasn't a compound of sulphur.

The cave's name derived from locals showing tourists the effects of the gas lake on dogs. Standing at normal adult height, humans were unaffected. Taylor, who later in his career objected to unnecessary animal experiments, watched as a dog held down in the gas lake 'made violent attempts to escape' because it couldn't breathe. When released from the cave, the dog began to recover. Taylor noted that, unsurprisingly, 'there was no individual present willing to undergo the experiment, to the same extent, on his own person,' so he lowered his face to the surface of the gas and noticed how it started to affect his sinuses. When he wrote about his visit a few years later, he accompanied it with a diagram showing a man in top hat and swallowtail coat, the level of the gas coming up to mid-thigh; this is presumably his self-portrait.

He arrived back in London in the winter of 1829. In March 1830, he became a member of the Royal College of Surgeons by examination, and he returned to France again in the summer, just as revolution tore through Paris. When Parisians hurled paving slabs and flower pots at the soldiers who shot at them, Taylor was at the Hôpital de la Pitié. He had a chance, unusual for a British student at the time, to see gunshot wounds and their treatment on a large scale.

In London again, Taylor set up in general practice, but this stage of his career lasted only a short while. The Society of Apothecaries were demanding improved education in the neglected field of medical jurisprudence, and Taylor was ready in the wings.

Unrecognised by any

Scotland had created the first British chair in Medical Jurisprudence and Police ('police' meant 'public health') in 1807 at the University of Edinburgh. As far back as 1789, lectures had been given there on the subject by Andrew Duncan. He had tried to convince the university of the subject's importance 'to every medical practitioner, who is liable to be called upon to illustrate any question comprehended under it before a court of justice'. The fate of the accused depended on the words of the medical professional, whose reputation was also on trial.

The adversarial justice system was developing, so that increasingly prosecution and defence would have counsel, and could employ expert witnesses. Counsel was usually a quick-witted barrister ready to unpick a witness's words or run them down with a rhetorical juggernaut: 'How cautious must, then, a medical practitioner be, when examined before such men, when it is their duty to expose his errors, and to magnify his uncertainties, till his evidence seem contradictory and absurd?' The first professor of Medical Jurisprudence at Edinburgh was Duncan's son; it did little to advance the subject in Britain, hindered because it was only an elective part of the course.

In 1822, 25-year-old Dr Robert Christison took the chair. Like Taylor, his background was in medicine and chemistry, and he had spent some time in Paris. He had read Orfila and attended one of his lectures. In 1823, he published a paper on poisoning by oxalic acid with his colleague Jean-François Coindet, and he began to be called into court as an expert witness by defence counsels. His lectures were poorly attended, and attracted mainly lawyers rather than medical students. From an initial class size of twelve students, by 1825 numbers had dwindled down to only one. That same year, Christison petitioned the university to include his course as an optional subject in medical degrees.

Christison became well known after he was a medical witness at the 1828 trial of Burke and Hare. Helping medical students gain access to cadavers, they had avoided digging up the dead and had taken to murder instead. Christison's opinion was sought on wounds and bruises; had they been given before death, during the act of killing, or after death due to the rough handling of the corpse? As part of his research to find out how the body bruised after death, Christison administered blows to corpses; a practical and yet ghoulish experiment, which nearly sixty years later, another Edinburgh Medical School graduate, Sir Arthur Conan Doyle, had Sherlock Holmes carry out in *A Study in Scarlet*. In 1829, Christison's *A Treatise on Poisons* was published and quickly became a standard text on toxicology. It had gone through four editions by the time of its last in 1845.

The University of London caught up with the academics north of the border, establishing the first chair of medical jurisprudence in England in 1828. Demonstrating the influence that the Scottish had in the field, the first professor, John Gordon Smith, was an Edinburgh graduate. He had served as an army surgeon until 1815, when he moved to London, but restrictions prevented him from practising there because his MD was from a Scottish university. In 1821, his book *The Principles of Forensic Medicine* was published, and in 1825 and 1826, he lectured on the subject, which had made him an ideal candidate when the university created its chair.

But the course wasn't compulsory; few were attracted by an elective course, and it wasn't funded beyond the fees of the small number of students who attended. Smith felt that it was 'unrecognised by any of the medical authorities in this kingdom', so he devoted himself with great energy to petitions, hoping to change the perception of his specialism.

In the late 1820s, Smith was knocked back from his application to be a coroner, being told that candidates would not be confined to those from 'the Legal, Medical, or any particular profession, but that any Gentleman of experience, respectable Character and liberal Education, is duly qualified to fill that office'. Smith vehemently disagreed, but his was almost a lone voice of protest; medical expertise was little valued in the legal sphere.

Thomas Wakley, a surgeon who had established the well-known medical journal *The Lancet*, was also turned down as a coroner. In the summer of 1830, he covered London with bill posters about medical jurisprudence and in August there was a public meeting on the Strand. Smith claimed 50,000 people attended, which seems an exaggeration, but the subject was very much in the public consciousness, following controversy over the death of a consumptive woman who had been treated by a quack doctor.

That September, the Society of Apothecaries made an announcement: from January 1831, in order to become a licentiate, candidates needed to have a certificate proving attendance on a three-month medical jurisprudence course. The course wasn't long, and it didn't demand an exam, but for the first time medical professionals would have to learn about the subject. It was too late for Smith. Despite giving an introductory lecture in October 1830, celebrating the fact that his campaign had won out, just a month later he resigned his chair, as he felt he was being snubbed by the university. Things came to a head when he found a book that he had donated to the library on a rubbish heap, and he took to drink. He died in a debtors' prison in 1833.

The peace and happiness of society

Anthony Todd Thomson, Professor of Materia Medica and Therapeutics at the University of London, presented the introductory lecture of his medical jurisprudence course on 7 January 1831, teaching alongside Andrew Amos, Professor of Law. It was the first medical jurisprudence course in England.

The Royal College of Physicians' library has a copy of Thomson's fifteen-page syllabus, which once belonged to Taylor. Given his early interest in the subject, it's likely that Taylor attended at least some of the lectures.

Thomson opened by emphasising the importance of the subject; correct medical evidence is a necessity 'to the cause of Justice; consequently to the peace and happiness of Society'. Thomson's part of the course was broken down into three areas: social relationships, the constitution of society, and personal safety. From rather gentle subjects such as how the voice is altered by age, there are issues such as how puberty is determined, and the 'period at which pregnancy is possible, and beyond which it cannot occur', how to identify hermaphrodites, the causes of impotency in males, and how medical professionals were to intervene in cases of sexual assault.

From pregnancy, he moved on to birth, where he addressed the question of the 'degree of deformity which constitutes a monster' and from thence on to infanticide. As abortion was illegal in Britain, but procured by various means including poison, medical practitioners needed to be able to tell abortion apart from natural miscarriage. How to tell if a newborn baby had died at birth or at its parents' hands was also on the syllabus.

'The constitution of society' looked at issues such as life insurance and legitimacy. Medical men were asked for their opinion on people's health by assurance companies so students learnt about 'diseases tending to shorten life, and which, consequently, affect the granting of policies of assurance'. Inheritance could be contested if a child was born prematurely; it might seem as if it had been conceived before the parents' marriage and hence its paternity could be in doubt. Seventy years before blood type analysis and well before DNA testing, Thomson discussed 'evidences of paternity from family likeness, colour, &c.'

Then there were the types of illness, disease and mental competency that would lead to disqualification from military and jury service. 'Soldiers often feign diseases,' he told his students. On to 'Mental alienation', where Thomson defined the sound mind, and identified the various unsound renderings of it, from epilepsy to intoxication, hysteria, 'nostalgia' and 'imbecility from old age'. He briefly covered public health, such as 'nuisances' (unpleasant or non-existent sewerage systems), factories, abattoirs, infectious diseases and cordons.

The third and last part of Thomson's course was personal safety. Beside his section on 'Injuries and Mutilations' ('slight – dangerous – mortal'), someone, perhaps Taylor, has drawn an asterisk. This section included 'the delay in obtaining surgical aid – the want of skill of the Surgeon employed' – issues reminding medical students of their professional responsibilities. Was the surgeon incompetent? Could this person have survived if medical help had been sent for sooner? They studied sudden death, and learned how to distinguish it from 'trance or catalepsy – suspended animation – feigned death'. They were also taught the 'mode of conducting the dissection of the body to ascertain the causes of sudden death'; the local medical man would be expected to conduct a post-mortem.

On dealing with 'persons found dead', they were taught that 'the causes of death are to be ascertained from the condition and state of the body,' and the place it was found should inform their investigation. They had to decide whether cases of hanging were suicide or homicide; could someone strangle themselves to death? Could you tell if a drowning victim was murdered or had taken their own life? If someone had starved to death, was it 'voluntary or forced'? Then there was smothering, wounds, and burns, which included the mundane difference between burns and scalds, the slightly more intriguing 'scorching from lightning', and the still contentious subject of spontaneous human combustion.

By far the biggest part of the section on 'personal safety' was on poisoning, taking up a third of the entire pamphlet. Thomson looked at how to find evidence of poisoning from examination of the body, and 'fallacies in judging of these arising from disease' – the symptoms of poisoning, such as vomiting and diarrhoea, can easily be the result of innocuous stomach bugs. He talked through organic and non-organic poisons, even 'the bites of venomous serpents'. The section under 'septic poisons' demonstrates how areas of scientific knowledge were still languishing in the doldrums. 'Septic poisons' referred to 'marsh miasma and putrefying animal and vegetable matter'. The 'bad air' theory of disease would be challenged by scientists later in the nineteenth century.

As Professor of Law, Amos's lectures mirrored those of Thomson's, examining his topics from a legal perspective. He took his students through practical courtroom issues, such as how 'medical men' were summoned to give testimony, and what expenses they could claim, as well as different sorts of evidence: hearsay or presumptive. Amos commented on specific cases, and when he moved on to the issues surrounding life assurances, he talked about 'exception in policies of suicide, duelling, or the hand of justice'. Pistols at dawn were not unusual in Georgian Britain. Amos finished his part of the course with perhaps his most useful lecture of all: 'enquiries usually made of medical witnesses on trial relating to violent deaths'.

The Hags of Rhubarb Hall

It was a lot to absorb in three months, and without an exam at the end, the more baroque gyrations of the law could slip out of the students' grasps; but it was an important start, even if *The Lancet* did not agree. 'Fudge; mere fudge!' they declared in a leader titled 'The Fudgery of Medical Jurisprudence'. Despite the journal's editor being so enthusiastic about medical jurisprudence, their excoriating leader a week later heaped insult upon the Society of Apothecaries, on Thomson and on Amos.

They called the Apothecaries 'the Hags of Rhubarb Hall'; referring to Apothecaries' Hall, this double-edged insult gendered them as women and mocked rhubarb, used well into the nineteenth century to cure various ailments.

They claimed that Thomson's 'materials are altogether crude and undigested, and are arranged in the very worst manner' and that his language was 'execrable'. *The Lancet* had long complained about the Apothecaries' certificate system for their licentiates, which they said was 'a barefaced and unqualified practice of extortion', and they claimed that the Apothecaries had moved their own goal posts so that professors at the University of London could teach courses for their licentiates.

Surgeons were still excluded from teaching would-be apothecaries, and the article makes an extreme comparison with the Corn Laws: 'This tax upon the stomachs of the poor for the benefit of the landowners exactly resembles the tax upon the minds of medical students made in favour of the fellows and licentiates of the Royal College of Physicians.' They objected to the idea of medical jurisprudence as a 'distinct branch of science', claiming that there was nothing wrong with physiological and toxicological teaching, except that 'the medical colleges and companies have been the only barriers opposed to a greater degree of perfection.'

But a few months later, at the start of the autumn term, 1831, many of the London hospitals began their lectures in medical jurisprudence, aimed at apprentice surgeons as well as apothecaries. Taylor, with his experience at the Continental hospitals, and his early interest in the subject, was an obvious candidate for the post at Guy's. He gave six probationary lectures, on poisoning, wounds, child murder and insanity, before his successful election to the chair. Aged just 24, he was now Professor Taylor, a title by which he would be known for the rest of his life. 'Professor' did not imply what it does now; it meant a lectureship, rather than high academic rank and a lavishly funded chair, but it did confer on Taylor an aura of wisdom.

Despite his youth, he was a striking figure at the lectern, tall and imposing. A portrait taken a few years later shows a young man of alert intelligence, with a shock of thick, dark hair brushed back from a wide forehead, his lips pursed in a knowing smile. His aquiline nose gave him a haughty air, and his large, protuberant blue eyes implied that nothing passed him by. [Plate 10]

The adverts that Guy's Hospital placed in local newspapers across England show what a draw the London hospitals were. The list of courses was headed by Dr Bright, after whom Bright's Disease is named, and Dr Annison, with their lectures on the Theory and Practice of Medicine. Eleven other courses are mentioned before Mr A. Taylor appears at the end of the list, one of the youngest lecturers at the hospital.

Chapter 2

More of Impulse than Discretion

1831–33

The cultivation of the science

From 1830 to 1832, Taylor worked in general practice, based in his rooms at 35 Great Marlborough Street in Soho. The street runs east from Regent Street, parallel to Oxford Street, and in the eighteenth century had been very fashionable. It was a sensible address for a young medical practitioner, even if Guy's Hospital was 3 miles away.

Continuing where he had left off with his Italian journal articles, Taylor started to write again. From 1831 to 1833, he wrote many articles for the *London Medical and Physical Journal*, showing the breadth of his interests and knowledge. He kept on the alert for cases that demonstrated the need for the study of medical jurisprudence.

At the 1832 Lent Assizes in Huntingdon, a gaoler called Henry Russell found himself on the other side of the prison bars when the woman he had got pregnant died after taking the arsenic he had given her to bring on a miscarriage. Russell was a married man, so her pregnancy was highly inconvenient.

Russell's case raised difficult legal questions; although a jury found him guilty and the judge passed the death sentence, it was respited to allow time for the judges to debate it. The legal issue needed to be resolved as cases like this weren't going away. Fifteen judges couldn't agree whether or not Sarah had committed suicide, and although they were unanimous that Russell was an accessory before the fact, they couldn't agree on the Act that he would be tried under.

This was just the sort of case to attract Taylor's attention, a perfect example of medicine colliding with the law. Unfortunately, in his enthusiasm, Taylor aired his opinion before the judges had met. He was pulled up by the *Morning Herald* for a letter he had written to the *English Chronicle and Whitehall Evening Post*, who said, 'We would recommend to this gentleman to be less precipitate for the future to engage in legal controversy in a way that betrays more of impulse than discretion, and to be sure, before he opposes an argument, that he understands it.' It was mortifying, perhaps all the more so because the *Morning Herald* had no idea who this correspondent was, 'a Mr Alfred S. Taylor, "Professor of Medical Jurisprudence and Medical Police in Guy's Hospital"' – the quotation marks sound rather scornful.

In July, Taylor and two colleagues from Guy's and St Thomas's were hired by the Rock Insurance Office to give evidence in the case of an apparent suicide. Mr Kinnear had died suddenly the year before, found with three pints of blood in his stomach. He was a banker, and had insured his life for several thousand pounds. If it could be proved that he had committed suicide, then the Rock wouldn't have to pay the insurance money to his widow. Taylor and his colleagues had to base their opinions entirely on the findings of an unimpressive post-mortem, and were unable to analyse the viscera.

Taylor was certain that death wasn't caused by a burst blood vessel in the stomach, and he said that from the circumstances around the death 'he would infer that some narcotic agent had been the cause of Mr Kinnear's dissolution.' But under cross-examination, he said that he hadn't known three pints of blood to be found in the stomach of someone who had died of poisoning. So what did Taylor mean about the circumstances of Kinnear's death – especially when Kinnear had walked across the room and rung a bell two hours before he died, which seems impossible if he had taken a lethal dose of laudanum several hours before? The experts were cornered; they were being paid by the insurance company, and without having seen the body, anything they said would sound awkward and uninformed.

The Lancet was mocking. The trial gave the readers 'a fund of amusement'. All the medical witnesses came under attack, Taylor targeted for his youth; he was 'the young lecturer … learned but somewhat beardless'. They 'appeared in the witness box more like interested advocates of the insurance company than impartial commentators on the facts before them.' This would be a criticism often faced by scientific witnesses in the courtroom. Taylor's career was off to an inauspicious start.

That same year, Taylor wrote five articles for the *London Medical and Physical Journal*. His first, in their June issue, was on 'Poisoning by Sulphuric Acid'. 'Perhaps,' he wrote, 'it will prove interesting to those of your readers, whose attention has been directed to the cultivation of the science of Medical Jurisprudence.' He described the symptoms endured by 19-year-old servant Isabella Creswell, whom Taylor anonymised as M- C-. His friend John George French, identifiable from newspaper reports, surgeon at St James's Infirmary in Soho, was suspicious of her symptoms on admission.

Isabella would not tell French what she had taken, so he went to the house she had been brought from to investigate. The parts of her bedclothes 'which were coloured *blue* had become *red*, and upon applying the tongue, it was found that the wet portions were acid.' Testing the soiled bedclothes of a dying woman with one's tongue is surely above and beyond the call of duty, but if French could find out what she had taken, he had the chance to save his patient's life. Even so, there wasn't much they could do other than administer magnesia, and hopeless treatments such as leeches and 'fomentations' (poultices). She finally admitted to having taken 'common vitriol' (sulphuric acid) but would not say why.

French's exertions were in vain; the girl died four hours after admission. A post-mortem showed how much internal damage the acid had caused. French called on his friend's laboratory skills, so the two men returned to Isabella's room and Taylor gathered evidence. He noticed that her blue cotton gown 'was tinged of a bright red wherever the fluids ejected had fallen' and how the bed linen was similarly bleached. He cut out samples from the counterpane to test, and results showed the presence of sulphuric acid. He tried the same test with her stomach contents, but could not find the acid, yet he was satisfied that 'the statement of the deceased was borne out by the result of the chemical analysis'.

Taylor concluded his article by explaining why this case is 'worthy of the notice of the medical jurist'. Firstly, the medical practitioner who had first been called to attend Isabella should have realised she had ingested poison. Her previous good health the night before, coupled with the colour change of the linens, and the condition of her tongue and throat, should have raised suspicions before she was sent to the infirmary. Secondly, Taylor knew he might not have been able to find acid in the stomach, hence he had taken a sample of the linen to test. He explained that mineral acids behave differently in the body from poisons such as arsenic or mercury, and sulphuric acid is fluid, easily ejected by vomiting.

This early article, written when Taylor was not yet 26, reads much like his later texts and reports of his scientific evidence given at trials and inquests. The readership of the *London Medical and Physical Journal* did not need him to explain where in the anatomy one might find the scrobiculus cordis, but his text is not overwritten or unnecessarily verbose. He chose lucidity, and this has to be one of the reasons why the books he later wrote were on the shelves of professional and layman alike.

The following month's edition carried a report by Taylor about the death of the matron of St James's Infirmary from cholera. Living just round the corner, Taylor was kept abreast of events that occurred there by French.

At the end of 1831, a cholera epidemic had landed on Britain's shores; by 1832, it had arrived in London. It tended to be called 'Asiatic cholera', and is what we would recognise now as the disease that causes dramatic diarrhoea leading to 'rice water stools' and kills by dehydration; untreated, it is often fatal. 'English cholera', on the other hand, featured vomiting and diarrhoea. It was rarely fatal, and seems to be a term used to cover a variety of gastric upsets, sometimes including dysentery. No one knew what caused cholera; it wouldn't be until the 1849 epidemic that surgeon and physician John Snow, a colleague of French's at the infirmary and an acquaintance of Taylor's, would put forward the idea that cholera was waterborne, which contradicted the miasma theory.

Taylor was abundantly aware of the arguments raging in the medical profession about how cholera spread, with scientists taking sides, either contagonist or anti-contagonist. He carefully described how Mrs Birke, the matron of St James's Infirmary, developed symptoms of cholera after sitting up with two cholera patients, one after the other. She died within eleven hours. As the area around the hospital was clear of the disease, Taylor reasoned that it could only have come into the hospital via the two cholera patients. She had not left the hospital and no one else on the staff had fallen ill. It could not have been carried on 'bad air' and was therefore the sort of scenario that the anti-contagonists wouldn't accept. Taylor's tone switched, writing like a barrister demolishing a flimsy alibi. Birke's case surely proved the anti-contagonists wrong – but without germ theory, he couldn't explain why.

His next article was a long piece on 'the Laws of England as it relates to the Crime of Foeticide'. Attitudes towards Taylor's specialism of medical jurisprudence had moved on since *The Lancet*'s furious leaders of the year before – or at least, so Taylor wished his readers to believe. 'Medical Jurisprudence is now so rapidly gaining ground as a branch of professional study, that any apology for the appearance of this paper in a Journal, the avowed object of which is to aid in the diffusion of every science, directly or indirectly connected with medicine, will perhaps be deemed superfluous.'

Unsurprisingly, he alluded to the Russell case, Taylor pointing out that 'few assizes pass without the occurrence of cases of this description'. The law was confusing as the punishment was less fierce – imprisonment or transportation rather than execution – if the foetus hadn't quickened; if the mother was unable to feel it move in the womb. This was based on a distinction in the Bible, as if the foetus was not strictly alive until it moved. But a woman who might be hanged if she had aborted a foetus that had quickened is unlikely to admit to the court that she could feel it moving. So how could medical professionals decide, and should the law be changed?

Taylor wrote that the distinction was irrelevant. 'No boundary can be established, no line can be drawn, no principle recognised, which tends to put a lower price upon human existence at one period of gestation than at another, when, by the laws of man, the rights of the individual are in theory equal.' He reminded his readers that a pregnant woman sentenced to death was spared her life until the birth of her child if it had quickened, but hanged right away if it had not, which 'would be to punish the future offspring for the crime committed by its mother'. A rare person indeed would condone abortion in a medical journal in 1832, or argue for the provision of safe, legal procedures to obviate accidental poisoning deaths in the first place.

The October issue of *The London Medical and Physical Journal* that year featured not one but two articles by Taylor, both on quirks of the Italian landscape that combined his interest in chemistry and geology. He recounted his trip to the Grotta del Cane and the experiments he carried out there, delighting in pointing out the errors of previous writers. As he believed the cave contained carbonic acid, he turned his geology article into a piece on medical jurisprudence. He explained the mechanism of suffocation by carbonic acid, and pointed out the difficulties of identifying it during post-mortem. Taylor referred to several scientists, such as Humphry Davy and Orfila; when mentioning Christison, he wrote of 'his admirable *Treatise on Poisons*', setting his cap at his celebrated contemporary. At the back of the issue, amongst books received and appointments and resignations at various hospitals, is an advert for Taylor's lectures.

The imprudent pursuit

Performing experiments in a cave of mysterious gas and licking the bed sheets of a dying woman are clearly dangerous and ill-advised pursuits; there were few safety measures in place for the nineteenth-century scientist. In early 1831, for instance, a medical student at Guy's had pricked his finger while 'dissecting a putrid corpse'. He suffered great pain and fever, and at his wit's end, committed suicide by taking prussic acid.

In September 1832, 28-year-old Alexander Barry, Lecturer in Chemistry at Guy's, blew himself up while performing an experiment in his rooms at Furnival's Inn. So violent was the explosion that initial reports assumed he had been using gunpowder, but the Royal Society, which had elected Barry as a Fellow only a few months earlier, said that he had been experimenting with compressed gases. 'The chambers took fire from the explosion; the window-frames and part of the wall of the back room were carried into Furnival's Inn Passage.' Rescuers were held back by the thick smoke emanating from the room, and when they managed to get in, they found Barry in his bedroom 'lying on the floor, weltering in his blood, with scarcely a rag on his back unscorched'. He had nearly blinded himself, and 'his right cheek was laid bare to the bone, his nose was shattered, his left hand was nearly divided, and several of his ribs on the left side were fractured.'

Despite these horrific injuries, *The Spectator* optimistically reported that 'Mr Barry is expected to recover': he did not. He lingered a month until his death on 8 October. The Royal Society shed few tears for him; Barry was a 'victim to the imprudent pursuit of his chemical inquiries'.

Aikin, Taylor's former tutor, had lectured with Barry. He had been in his post at Guy's since 1821, but he was not a particularly gifted lecturer. Although chemistry was an essential subject for medical students, his lectures usually descended into chaos. With what remained of Barry lying in Whitechapel's Quaker burial ground, a new lecturer was required to share Aikin's load, and that man was Alfred Swaine Taylor.

The extraordinary investigation

At the turn of the year, a rather unusual discovery was made in the dissection room at Guy's Hospital. In 1832, the Anatomy Act had been passed, which meant that bodies destined for a pauper's burial 'on the parish', if unclaimed by family or friends, could be sold to medical schools for dissection. The poor had little agency over their own bodies, especially after death. The Act was partly in answer to the scandal of the Burke and Hare trial, and the copycat London Burkers, a way to provide a plentiful supply of subjects for the anatomists without the help of murderers or Resurrection Men.

One Sunday morning in January 1833, the porter of Guy's Hospital received the bodies of two women from the parish beadle of St Margaret's Workhouse in Westminster. One of the women was Eliza Edwards, a 24-year-old actress fallen on hard times. She was originally from Dublin, and after the death of her father, a lieutenant colonel in the East India Company, she had been raised by an uncle. Aged 14, she had run away to Paris and had taken to the stage. For ten years, she had lived with a woman called Maria Edwards as her sister. She acted on the provincial stage but by 1830 she had drifted to London, where she was kept by a succession of men. In life, she had been an attractive woman, with long, glossy, light brown hair; men found her enchanting.

She developed a chest infection, and was attended by Dr Clutterbuck, who had been called in by Eliza's most recent gentleman friend. He could not save her, and Eliza expired. Her 'friend' did not pay for Eliza's funeral and her honorary sister could not afford it, so Eliza would have been buried by the parish if Guy's had not needed some females to dissect. Eliza's body was produced in readiness, but an inconsistency was discovered: Eliza had been assigned male at birth, and had lived as a woman since the age of 14, when she had run away from home.

Eyebrows were raised and Lord Melbourne, then Home Secretary, directed that a coroner's inquest should be opened. As was usual at the time, the inquest took place in a pub. This was Alfred Swaine Taylor's first appearance in the newspapers as a witness at an inquest. He had performed the post-mortem, reporting that Eliza's liver was the worse for wear, presenting 'that appearance seen in persons addicted to drinking, commonly called a drunkard's liver'; despite this, it was lung disease that had killed her. Taylor noticed that Eliza had plucked out her facial hair, and that her dress had a curious high collar on it, to conceal her throat.

A puzzled juror asked Taylor, 'Has the head of the deceased been separated from the body?'

'It has not,' Taylor replied.

The next questions that Taylor was asked, and the answers he gave, were not printed in the newspapers. They 'excited the greatest astonishment' but 'the evidence is unfit for publication.' These were Taylor's examinations of Eliza's anus, the condition of which, he later wrote, 'left no doubt of the abominable practices to which this individual had been addicted'. Sodomy was not something that *The Times* could print for digestion by their readers over the breakfast table.

One of the parish overseers praised the hospital for the respectful way in which Eliza's body had been handled; 'he should have no objection to give his own body over to dissection.' But the jury decided that Eliza deserved no more respect. One juror thundered that it was 'a most extraordinary case of depravity' and that 'the body ought not to be buried in the usual way, and that some proceeding should be adopted to warn others from such unparalleled acts of turpitude, which reflected the highest disgrace upon human nature.' For Eliza Edwards had transgressed the mores of nineteenth-century British society. In their eyes, she was a man passing as a woman prostitute, who had regularly enjoyed an illegal sexual act that at the time carried the death penalty. And in convincing so successfully as a woman, as far as they could see, she had lured unwitting men to commit an immoral crime.

Merry japes during Aikin's dull lectures were one thing for the medical students, but the prurient novelty of the investigations now at hand was entirely too much for them. They interrupted proceedings several times, and the jury rebuked them, saying, 'that such conduct was indecent'. While the jury went to consider their verdict, the students continued to misbehave, and 'called the jury a set of humbugs and jackasses'.

Taylor did not forget Eliza. He wrote a paper about her in *The London Medical and Physical Journal*, and under the heading of 'concealed sex' would still be writing about her over forty years later.

None of the medical students or their lecturers at Guy's had known, as they puzzled over Eliza Edwards, the secret of James Barry, who had been among their number twenty years earlier. Barry had gained an MD from Edinburgh, and had trained at Guy's under Sir Astley Cooper. He became a regimental surgeon and performed one of the first successful caesarean sections. Taylor met him in the late 1850s, and whilst he thought there was something unusual about Barry, nonetheless accepted him as a medical gentleman. Barry was promoted to deputy inspector general, and was respected, although equally feared, brandishing an oversized sword and spurs. When Barry died in 1865, it was discovered that he had been assigned female at birth: in his teens, he had begun to live as a man, entering a career that, had he lived as a woman, would have been entirely closed to him.

Chapter 3

Fearful and Wonderful
1834–38

My beloved is mine, and I am his

On 11 July 1834, Alfred Swaine Taylor got married. His bride was Caroline Cancellor, the 24-year-old daughter of stockbroker John Cancellor. She was the youngest of five children, the only daughter. Her maternal grandfather had been a captain in the East India Company, and her eldest brother was a barrister who became a Master of the Court of Common Pleas. Two of her brothers were stockbrokers, and the youngest ran a starch manufactory.

The Cancellors were originally from Cambridgeshire; Caroline's grandfather had moved to London in the 1750s. It is possibly Caroline's father whose name is found in lists of donors to various charitable causes, such as a fund for the widows and children of Nelson's fleet, and for the relief of the Irish poor. He didn't remarry following his wife's death two months after Caroline was born, and he lived with his family on Upper Gower Street in the London parish of St Pancras.

In his will, written when Caroline was 18, John Cancellor made sure that his children would be provided for by money invested in stocks and funds. He was careful to word his will to ensure that Caroline's inheritance could not be squandered by any future husband; such measures are standard in wills that date from a time when a woman's property automatically became her husband's on marriage. It would protect his daughter from the attentions of any cad who had their eye on a wealthy orphan, as well as ensure her an income should any future husband be bankrupted – a very real fear for the professional class in the nineteenth century. When he died in December 1831, John left his daughter well provided for.

Richard, Caroline's second eldest brother, had died a bachelor a few months before her marriage. He left her the lease of his house on Cambridge Place (now Chester Gate), on the eastern edge of Regent's Park, as well as its contents: 'household furniture, plate, linen, china, books, paintings and prints, wearing apparel, watches, jewels, trinkets and ornaments of the person, wine and other liquors, horses, carriages and harnesses'. Richard had been living in some style in one of John Nash's Georgian houses for the well-to-do, and leaving his house and its contents to Caroline implies that she had been living there with him. He also left Caroline the income of some stocks and shares, to add to those already left to her by her father. Among legacies bequeathed to his brothers, aunts and friends, Richard left £100 to Mary Ann Swinley of Brighton, 'my sister's intimate friend'.

Mary Ann was one of the witnesses to Taylor and Caroline's wedding, along with Caroline's eldest brother, and Taylor's brother, Silas. Perhaps Mary Ann became a surrogate sister for Caroline, as well as for Richard. He may have entertained ideas of marrying Mary Ann, but his death intervened.

Alfred Swaine Taylor moved from Great Marlborough Street, on the busy northern edge of Soho, to Cambridge Place. He would no longer be dropping in regularly on John George French at St James's Infirmary, but the two men appear to have remained friends. Happy at home with his wife, Taylor was later remembered as 'a man of quiet and domestic tastes', who was 'little seen either in the medical societies or in social medical intercourse'.

Although Cambridge Place is a short side street rather than a sweeping Georgian terrace with a view of the park, it was still an elegant address. In 1841, Taylor could count among his neighbours a barrister and a wealthy rum merchant. About 400 feet from the Taylors' front door, in the south-east of Regent's Park, was the Colosseum, a sixteen-sided building that housed a painted panorama of London. Just to the south, in Park Square, was a diorama based on Louis Daguerre's in Paris, where paintings and lights were used to create dramatic tableaux. The Zoological Gardens had opened in the north of Regent's Park in 1828, and in 1832, archery grounds opened to the south, which were flooded in winter for ice skating. Within a few years of Taylor's marriage, the Royal Botanic Society Gardens were established in the park.

Immediately behind Cambridge Place is Albany Street, running north to south. To the east were the areas designed by Nash for the working class and artisans who serviced the wealthier households bordering the park. Three hundred feet immediately east of Cambridge Place is Clarence Gardens, one of the last addresses of Eliza Edwards.

For man walketh in a vain shadow

A year after their wedding, the Taylors welcomed their first child to their home at 3 Cambridge Place. Their son, Richard Alfred, was named after Caroline's brother and the child's father; he was baptised at St Pancras church, where his parents had been married. Would this little boy grow up to follow his father into medicine or science, or would he take the route of his mother's family and become a stockbroker or barrister? They would never find out: Richard died only five months later, in January 1836. He was buried at St James's in Piccadilly, the parish church of Taylor's old home on Great Marlborough Street.

Despite this family tragedy, Taylor was busy. Firstly, he had reason to be glad, because 1836 saw the passing of the Medical Witnesses Act, which enshrined in law the principle that medical witnesses would be paid one guinea for attending an inquest to give evidence, and two guineas for performing a post-mortem, with or without analysis of the stomach or intestines. It showed that medical witnesses were increasingly valued.

The same year saw the publication of Taylor's first book: volume one of *Elements of Medical Jurisprudence*. Coming in at just over 500 pages, it cost the princely sum of fifteen shillings. *The London Medical and Surgical Journal* said it chiefly consisted 'of recent British cases of forensic medicine, and will be found a very valuable addition to the medico-legal library'.

The Brighton Gazette praised it highly: 'an ably written treatise on a most important, but strangely neglected subject'. Their reviewer argued that medical jurisprudence was of interest to people outside the profession, especially those who might one day serve on a jury. 'The gross fallacies that have been promulgated, the vulgar errors that still prevail on many points, are ably exposed in the volume before us,' they wrote. The reviewer was particularly impressed by Taylor's section on asphyxia, which they felt was 'a favourable specimen of the author's abilities'. They concluded, 'We are sure that no one will peruse these pages with indifference, who feels the slightest interest in understanding the physical laws that regulate the existence, and the modes in which external accident and violence disturb the functions of the "fearful and wonderful" frame of man.'

Unfortunately for Taylor, a new British edition of Beck's *Elements of Medical Jurisprudence* came out that year. It seems this overshadowed Taylor's production, which must have been frustrating, even though it was Beck who had initially fired his interest in the subject.

And so Taylor went on holiday, presumably taking Caroline with him, as he went to Brighton, home of her friend Mary Ann Swinley. Taylor's name appears in a list of visitors to the Sussex Scientific and Literary Institution and Mantellian Museum in August that year. His name is included with lords and ladies as well as other men of science, such as Richard Bright, his colleague from Guy's.

Taylor may have gone with Silas; in June the following year, Silas married Mary Ann Swinley in Brighton. Neither Taylor nor Caroline was a witness to this marriage, but a man called Charles Atkinson was. Charles's brother, William Atkinson, was Mary Ann's brother-in-law, and the Atkinsons were also cousins of the Swinleys. William Atkinson was a manufacturing chemist, so it is possible that Taylor knew him professionally, as well as through his family links; it could even explain how Taylor and Caroline had met.

This beautiful method of analysis

In 1836, Guy's Hospital began to publish its series of *Reports*, which were produced until 1974. Articles were contributed by hospital staff, reporting on cases that they had worked on, and Taylor was to become one of the more prolific contributors. He was not alone: Bright also wrote frequently, and about far more parts of the body than the kidneys. The journal was intended to be of benefit to the whole medical profession, not just for the hospital's staff and students.

Taylor's very first article in *Guy's Hospital Reports*, published in 1837, was on James Marsh's new test to detect arsenic. Marsh had published his findings the year before, in the *Edinburgh New Philosophical Journal*; that it was published in the city that had the first British chair of medical jurisprudence, and home of toxicologist Robert Christison, might not be a coincidence.

Before Marsh came up with his test, there had been various ways to identify arsenic, but none were particularly reliable. In Christison's 1832 edition of *A Treatise on Poisons*, he discussed an extraordinary number of tests for arsenic. He outright rejected the late eighteenth-century test that relied on the analyst's sense of smell; this method stated that arsenic gave off a garlic-like odour when burnt. Christison said this only applied to metallic arsenic, not arsenic oxide. It was the oxide that was most likely to be used as a poison, 'on account of the facility with which it may be procured in this country, even by the lowest of the vulgar, and the ease with which it may be secretly administered'.

Christison knew from personal experiment how easy it was to conceal it in food and drink. He had 'repeatedly made the trial, and seen it made at my request by several scientific friends' whereby they extended 'the poison along the tongue as far back as we thought safe' and 'we all agreed that it had hardly any taste at all – perhaps towards the close, a very faint sweetish taste.' Christison had put arsenic oxide in his own mouth to analyse its flavour, but it was so slight that it could be easily disguised. The second test he rejected was one in which the oxide is mixed with 'carbonaceous matter' and heated between two copper plates; it would produce a silver alloy, but needed at least a grain (about 65mg) to be effective.

Christison favoured the reduction test, heating the oxide in a tube, which would separate out the metal, making it visible to the naked eye. If the oxide was in a liquid, then hydrogen sulphide gas could be used to create arsenic trisulphide, and the hydrogen would react with the remaining oxygen to form water. Then the arsenic trisulphide could be reduced in the same way as the oxide. Other methods he suggested were the introduction of chemical reagents that could produce a precipitate. He suggested that three different precipitate tests were required, to cancel out any fallacies: hydrogen sulphide, ammoniacal silver nitrate, and ammoniacal copper sulphate. The same colour reaction could be produced by other substances, but only arsenic could produce the looked-for result by all three.

Finally, he explained how to perform a test for arsenic in organic fluids and solids, which was the most important test he outlined, 'for in nine cases out of ten the subject of analysis is the stomach with its contents.' The precipitate tests would not work, because they relied on the analyst noticing colour changes, which would be obscured by the pre-existing colour of stomach contents. Christison believed that his method was the best; firstly, he procured a transparent solution by boiling up the suspect material, and then filtered the liquid from the solids, which could take up to thirty-six hours. Various chemicals had to be used to remove 'animal matter' from the liquid, and then the hydrogen sulphide test followed. Christison believed that this was the foolproof way to find arsenic, without interference from 'animal matter'; he claimed he'd never failed to find arsenic in the most complex fluids. When Christison wrote about cutting up solids and adding distilled water to make it easier to boil, he wasn't only testing stomach contents or vomit: he was also cutting up human organs, in case poison had been absorbed into the body.

In 1833, James Marsh was the expert witness in the case of George Bodle, who, it was thought, had been poisoned. Marsh identified arsenic in the coffee that Bodle had drunk before his death, but his grandson, the prime suspect, was acquitted at the trial; later, however, he confessed. Dissatisfied with the standard tests, Marsh, who worked as a chemist with Michael Faraday at Woolwich Arsenal, decided he would come up with an improved method.

Marsh's idea was elegant in its simplicity. Rather than dig about trying to separate the arsenic from the liquid, he forced the arsenic to reveal itself. It was at once a way to separate out the arsenic and to test for its presence. Marsh used hydrogen gas, which deoxygenated the arsenic oxide; the arsenic would then combine with the hydrogen to create arsine (arsenic trihydride). Now turned into a gas, the arsine would separate itself from the liquid. Christison described it as 'beautiful', and said 'that injustice has been done its discoverer both by himself and those who have since investigated the subject, when they denominated it merely a test.' This was no test, but a *process*.

It required a U-shaped tube, into which a piece of zinc was added. The liquid to be examined was mixed with sulphuric acid and then poured into the tube. 'Bubbles of gas will soon be seen to rise from the zinc, which are pure hydrogen, if no arsenic be present; but if the liquor holds arsenic in any form in solution, the gas will be [arsine].' The gas was ignited, and a piece of cold glass held over it. Any arsenic present would be deposited in its metallic state on the glass. White arsenic could be produced by holding a glass tube over the burning jet of gas; becoming oxidised, the tube would be lined with arsenic oxide. Despite Christison's distrust of chemists' nostrils, Marsh still felt it worth mentioning that a garlic smell was noticeable. Once the arsenic had been extracted, then the reduction and precipitate tests could be performed.

Marsh could extract arsenic from samples of any size, and he could perform the test even when organic matter was causing the sample to froth inside the tube. He had managed to find 'one grain [65mg] of arsenic in 28,000 grains of water', and identified the arsenic in just one drop of Fowler's Solution, a tonic that was known to contain arsenic; this was only one-120th part of a grain. Marsh believed the process employed such simple equipment that there was 'no town or village' in which it could not be performed – an important point when any medical professional might be tasked with investigating a suspected poisoning.

Taylor didn't have to wait long to try out Marsh's test. On 17 May 1836, a 25-year-old woman was admitted to Guy's; she said she had taken arsenic. In his article for *Guy's Hospital Reports*, Taylor described her symptoms and the appearance of her organs at the post-mortem. He commented on the unusual fact that the woman did not report any pain, but he said that Christison and Orfila had come across similar cases; 'we ought, therefore, in suspected cases, to be prepared for such an anomaly.' It was vital that anyone called in to pronounce on a suspicious death didn't fall into the trap of expecting all poisoning symptoms to be uniform.

The patient experienced extreme thirst, something Taylor had noticed 'in three other cases which have fallen under my observation', and he remarked how strange it was that British toxicologists gave the symptom little attention, whereas his Continental colleagues did. He commented on the length of time that it took the patient to die, an important point to mention because the speed with which someone could die of arsenic poisoning might be contested at a trial, the belief holding sway that a speedy death from the poison wasn't possible. Taylor and Christison both asserted that it was, sometimes in as few as three hours.

It was the perfect case for Taylor's first attempt at the Marsh test: 'The confession made by the deceased, as well as the nature of the symptoms, left but little doubt that arsenic had been taken: but a case is always rendered more satisfactory when, to these sources of evidence, we can add the certainty commonly derivable from chemical analysis.'

He divided the stomach contents into two parts. He performed Christison's previously favoured test on the first part, and found that it tested positive for arsenic. Then he used 'the ingenious apparatus which had been then but recently proposed' by Marsh, and found arsenic too.

A few months later, on 3 September 1836, another woman was admitted to Guy's, having taken arsenic. Taylor didn't name her, only giving her age and saying that she had taken 'an ounce of arsenious acid, in consequence of some disappointment', but newspaper reports suggest that she was Ellen Dickson of Southwark. She had been married only two months when her husband left her for another woman. Taylor outlined the symptoms again, with the vomiting and diarrhoea so characteristic of the poison, the body doing its best to rid itself of the irritant.

Ellen had admitted to one of her fellow lodgers that she had taken arsenic, and a surgeon was sent for; she resisted his help and knocked the medicine he offered her out of his hand. She was taken to Guy's in a coach, and a stomach pump was used, but it was too late. No arsenic was found in her stomach at post-mortem, so once again seeing the chance to try Marsh's new process, Taylor tested some of Ellen's vomit and the liquid brought up by the stomach pump.

He performed two tests, one using black flux, which produced an almost imperceptible indication of arsenic being present. He analysed the remainder with Marsh's test, and found arsenic. Although the quantity obtained was small, 'it was sufficient to shew that arsenic was really present … which, without the application of Mr Marsh's apparatus, I do not think it would have been possible for an experimentalist to have declared.'

Taylor was impressed, and took to task the objections that had been levelled at Marsh's test. While it was possible that sulphuric acid and zinc, which are required for the Marsh test, could contain arsenic, all one had to do was to test samples of it in advance for their purity. And if they were tainted, 'purer materials must be sought for; and these will not commonly be difficult to find.' It was true that other substances could combine with hydrogen and, when burnt, form a deposit on glass. But in exasperation Taylor said, 'The plain and obvious answer to this objection' was that one simply analysed the residue using the standard arsenic tests. He had spent six months experimenting with Marsh's innovation, 'both on artificial mixtures and on the contents of stomachs of persons poisoned by arsenic', and he was impressed by its sensitivity: he had obtained 10mg of arsenic from 45,000 parts of water.

After his review of Marsh's process, Taylor's article went on at some length remarking on arsenic's solubility, an issue that scientists did not agree on. His article shows that he read widely, from scientists writing in French and German as well as in English, and that he saw the legal importance of his enquiry: arsenic's solubility had an impact on trial outcomes.

His article was praised by the *British and Foreign Medical Review*; it was a valuable paper, recommended to anyone interested in the medico-legal issues surrounding arsenic. Points likely to come up at a poisoning trial were 'settled on the only true basis – that of cautious induction from well-conceived and carefully executed experiments.'

Taylor's renown was growing: in 1838, he temporarily stood in to give lectures in forensic medicine at King's College while their chair was vacant; in the same year, he wrote four articles for *Guy's Hospital Reports*. But something was about to grab his attention, something that would combine his expertise in chemistry not with medicine, but with art.

Chapter 4

The Light of an English Sun
1839–41

Drawings of shadows

William Henry Fox Talbot was one of photography's pioneers in Britain. Chemists knew that under certain conditions silver reacted to sunlight, so Talbot, frustrated with his inability to sketch views of Lake Como in Italy, came up with an idea. He coated writing paper with table salt and silver nitrate, and placed an object on the paper. He left it in sunlight, so that the uncovered paper would darken. When the object was removed, it left a perfect silhouette on the paper, a negative that Talbot called 'sciagraphs' or 'drawings of shadows'.

Talbot continued his experiments, and sciagraphs became sensitive enough to be used in camera obscuras, the wooden boxes that were early cameras. But he was distracted by other things, only returning to photography in November 1838, when he began to write up his experiments for the Royal Society. When news of Louis Daguerre's discovery arrived in January 1839, Talbot was in a quandary; there were no details revealed of Daguerre's method, and if Talbot's process was the same, then he would lose all claim to his discovery. On 25 January, Michael Faraday showed some of Talbot's sciagraphs to the Royal Society, and read Talbot's paper describing his process. It was called 'Some account of the Art of Photogenic Drawing, or the Process by which Natural Objects may be made to delineate themselves without the aid of the artist's pencil'.

Alfred Swaine Taylor started his experiments at once, trying to replicate Talbot's results based on 'a short abstract of Mr Talbot's method' in *The Athenaeum* journal. Not having any luck, he came up with his own innovation.

Talbot's process used paper, which Taylor saw as having advantages over Daguerre's plate of silver. Writing in 1840, Taylor confidently – and wrongly – predicted that 'the process of M. Daguerre is so complex, expensive, and requires so much practice, that even had it not been "patented", it is not likely it would have ever come into general use.'

Talbot's method was to soak paper in a weak solution of salt, and paint a strong solution of silver nitrate over it, thus creating silver chloride in the paper. Taylor found this didn't work, 'from the chloride not being equally formed in it'. So in

February 1839 – only a couple of weeks after he had initially heard of Talbot's discovery – Taylor came up with a plan, utterly characteristic for a toxicologist. He hit on the idea of using ammoniacal silver nitrate, one of the three reagents used in precipitate tests for arsenic.

After treating the paper by painting it on one side, Taylor dried it 'in a dark drawer or closet'. He spent months trialling his method, apparently at home. It's tempting to imagine his wife or one of the servants opening a cupboard at 3 Cambridge Place in the course of their diurnal housework, only to find that Taylor had stuffed it full of his specially treated paper.

He experimented with different items: leaves and flowers, feathers and lace. 'Perhaps no objects have ever been transferred to paper with the truth of nature so firmly stamped on them as feathers.' Light passed through white feathers too easily, but Taylor could copy the feather of an Argus pheasant in an exposure of twenty minutes. 'The overlapping parts, as well as the fine down around the barrel, is often so delineated as to give the appearance of the feather itself being actually transferred to the paper.'

By letting light fall through the reverse side of a sheet of paper, he found he could make what are, essentially, very early photocopies, albeit in negative. Emerging from a culture where copying was entirely down to the skill of a draftsman, Taylor's enthusiasm for what he was now able to do through practical application of his chemical knowledge is palpable.

As with white feathers, white lace would not work, but a piece of black lace could be copied in a couple of minutes. Examples of Taylor's photogenic drawings have survived in two albums, which are in private hands. In an album that belonged to Ellen Shaw, a family friend, there is an image captioned 'A piece of old lace of Mrs Taylor's, I put on for Dr Taylor to put on the top of his house in the sun'. It's not clear when this image was made, although if the title 'Dr' is correct, it would be from the late 1840s at least.

The image is haunting; a piece of black lace worn by Caroline Taylor, perhaps in mourning, turned ghostly white by the action of chemicals and sunlight. The detail of the lace and its fine threads can be clearly followed, the broken strands easily made out. The caption reveals Taylor on the top of his house, perhaps the roof, where he could best capture the sun's rays. And it also shows that he did not close science away from women.

The process didn't end here. 'Unless there were some means of permanently fixing the impression, when once taken, the art of photogenic drawing would merely serve to gratify a temporary curiosity.' To be able to view the images in sunlight, they had to be fixed somehow to prevent the sun continuing to act on the chemicals, obliterating the image. Taylor tried several chemical combinations without success, before he remembered Sir John Herschel's work with hyposulphites. The best method, he decided, was to pour near-boiling water over the image to wash it clean of the remaining silver, and then treat it with hyposulphite of soda or lime.

I make no secret of my process

In April 1839, although he still had some way to go with his experiments, Taylor explained his process to Michael Faraday, and loaned him some of his photogenic drawings for an exhibition at the Royal Institution. 'I make no secret of my process,' Taylor proudly told him. In the summer of 1840, Taylor self-published his research in a pamphlet, *On the Art of Photogenic Drawing*. He wrote that he had been experimenting for sixteen months, and he lent some of his images to *The Athenaeum* journal to accompany their review of his pamphlet; they were impressed.

Perhaps it was Faraday who had voiced concerns about Taylor's choice of chemicals, as Taylor had to explain in his pamphlet that it wasn't dangerous. A similar compound, ammoniuret of silver, was known to be explosive, but worry not, as 'it is made by a process totally different to that here recommended' and the medical profession had used ammoniacal silver nitrate safely for nearly thirty years in arsenic testing. Considering the fate of Alexander Barry, one assumes that Taylor took care not to go a similarly explosive way.

Taylor's pamphlet is written so clearly that even today you could follow his instructions and make your own photogenic drawings. That said, the stationers and chemists he recommends as purveyors of the finest papers and purest chemicals ceased trading long ago. Taylor never patronised his readers, using his pamphlet as a way to share information and, to an extent, teach the curious in the ways of chemistry. When explaining the various problems he faced in fixing the image, he writes: 'To readers not acquainted with chemistry, it may appear singular, that while a diluted solution of hyposulphite of soda destroyed the drawing immediately, a concentrated solution preserved it for many weeks.' He explained the chemical process that causes this counter-intuitive result, so that readers come away with a far better understanding of the process than if Taylor had merely written a set of how-to instructions. It says much for how he came across to his students at Guy's, and how he was able to explain scientific evidence to a jury who may never have seen a test tube in their lives.

Taylor objected to patents and would not patent his own process. When discussing successful experiments with photogenic drawings on ivory, he made a sarcastic comment about Daguerre: 'Supposing the light of an English sun not to be included in the patent of M. Daguerre, there seems to be no objection to the use of these ivory plates in the camera.'

Although Talbot moved on to make calotypes with camera obscuras, and Herschel invented the cyanotype in 1842, which was a much quicker way to produce shadow drawings, Taylor's work was not insignificant. In 1842, he helped Henry Collen, Talbot's first licensed photographer, and in return, Taylor examined Collen's camera so he could have one made for himself. Still sniping about patents, Taylor wrote to Collen, 'I certainly shall take care to keep it out of the patent clutches of Mr Fox Talbot.' Robert Hunt, in his 1853 *A Manual of Photography*, wrote that photography was indebted to Taylor.

The two surviving albums of Taylor's images contain his photogenic drawings, as well as several sketches made on his Continental travels. There is a photograph showing the higgledy-piggledy back elevations of houses, possibly a view from the window across his back garden. He drew a cartoon sketch of a travelling photographer; on the side of the waggon Taylor wrote, 'Cambridge Photography Saloon' and 'Portraits taken in all weathers'.

Filii Morientes Posuerunt

On 28 September 1840, Taylor's father died in Northfleet. He was 69 years old, and was buried with his first wife. At the foot of the headstone it says, *Filii Morientes Posuerunt* – erected by the sons of the deceased. The memorial boasts of Thomas Taylor's role in the East India Company, and declared his Norfolk origins, but does not mention his long career as a merchant.

Although Oxley and Taylor was trading from 8 George Yard, off Lombard Street in the City of London, by 1836 Thomas advertised as trading from his home at 1 Granby Place on the High Street in Northfleet. [Plate 9] Granby Place still stands today; two Grade II listed semi-detached Georgian houses, not far from the church.

Thomas left the majority of his estate to his sons, with legacies to his two surviving siblings. A month before his death, he added a codicil, allowing his second wife, Elizabeth, to live at Granby Place, with an annuity of £40. After her death, Granby Place and the annuity would pass to Taylor and Silas.

The codicil mentions two grandchildren. Taylor and Caroline had not had any more children following Richard's death, but Silas had two children, Emily and Basil. Basil was only a few months old and was left shares in a bank. Two-year-old Emily received shares in the Abney Park Cemetery Company, one of the 'magnificent seven' garden cemeteries opened in the mid-nineteenth century to cope with London's spiralling population and crowded churchyards; an intriguing legacy for a toddler.

A snapshot

On the night of Sunday, 6 June 1841, the census was taken in Great Britain. A snapshot of the country on that night, it was the first British census to record the name of everyone, rather than just the head of the household. Visiting a friend, staying in a hotel, or sleeping in a barn to follow the hay harvest, your name and age might appear, with varying degrees of accuracy.

Taylor and Caroline were recorded that night at Cambridge Place, with four female servants. Merchants, barristers, medical men, and people of independent means lived in the houses around them. A mile and a half away on Regent Square in Bloomsbury, Taylor's brother Silas was living with Mary Ann and their two children. [Plate 10] Their neighbours were similarly placed in society to Taylor's: barristers, auctioneers, even a vocalist. Today, only one side of Regent Square would still be recognisable to Silas, with its five-storey Georgian terrace; the other three sides were destroyed in the Blitz.

The huge variety of life recorded in the census, the teeming millions of Great Britain, were all, Taylor feared, at risk. An invisible menace was amongst them, a colourless powder that had almost no taste, but, thanks to Marsh's innovation, could be detected. There were many poisons that might be used to take away a life, but it was arsenic that Taylor feared most, and it would kill by accident or design.

So frequent in the present day

In October 1839, a friend of Taylor's had arrived at his door with a bottle of port wine: not a gift, but a dangerous artefact. The evening before, Taylor's friend had been at a dinner party, where three people who had drunk from the bottle had displayed symptoms of poisoning; luckily none of them had died. Taylor found that the wine contained arsenic, both dissolved in the liquid and in sediment at the bottom of the bottle. But how had it got there?

Taylor wondered if lead shot had been kept in the bottle before it was used for wine; arsenic was used in its manufacturing process. He began an experiment in December 1839 that would last over a year, until January 1841. He put shot in two different bottles of wine, and the experiment showed that arsenic didn't leach out of the shot in large enough quantities to explain that found in the bottle delivered by the dinner party guest. The only other reason Taylor could think of was that the bottle had been reused. It could have been sold to a pedlar who hadn't realised that it had previously contained arsenic, and then had sold it on to a wine merchant who hadn't washed it properly. Taylor directed that bottles that had once stored arsenic should be destroyed.

He described a recent case from Jersey, when a family of four all succumbed to the symptoms of arsenic poisoning after drinking perry from a stoneware jug; one of them had died. The doctor investigating the death smashed open the jug and found a deposit of arsenic inside. Research into the jug's provenance showed that it had 'at some previous time, [been] a receptacle for bug-poison'.

This reuse of bottles was something to which, Taylor said, 'public attention has not been sufficiently drawn', because there was no vigorous system of public health. Taylor had been paying attention to the cases of poisoning that had happened in the last decade in England. He believed 'that many cases of this kind have occurred, especially with stone bottles such as those which are used for the sale of ginger-beer, in the streets of London. The illness or death has been ascribed to cholera, or some imprudence on the part of the affected person, when the real cause was probably some irritant poison.' If this sounds paranoid, it was Taylor who had tested the bottle of port and found arsenic in it; it was Taylor who was well aware of the different legitimate uses of arsenic, whether as rodent killer, sheep dip, or dye. And it was Taylor who had examined the bodies of people who found life so intolerable that they ended it by willingly swallowing poison. Arsenic was all around the average Victorian; it surrounded Taylor and besieged him.

When he included the dinner party poisoning in an article for *Guy's Hospital Reports* in 1841, it was to work out how much arsenic had been consumed by the dinner guests without killing them. Taylor believed he could estimate this from the amount of arsenic he found in the remaining wine. There was no consensus on how large or small a dose could be fatal, and Taylor pointed out that it was never a fixed amount: it depended on the victim's health and whether or not they had taken the arsenic on a full stomach. If they had recently eaten, as in the dinner party case, the arsenic would be harder for their body to absorb, and much more easily ejected by prodigious vomiting.

What do people expect to find?

In February 1841, John Breach, a former student of Taylor's, returned to Guy's, bringing with him the stomach of the late Cornelius Rhymes. Breach had seen Rhymes on 27 January because he had diarrhoea; he initially thought nothing of it as it was rife in the village of East Hagbourne, Berkshire, where Rhymes lived. Not considering him to be in any danger, Breach prescribed a simple medicine of chalk and opium.

Breach was surprised when his patient died two days later, and despite the symptoms being violent vomiting, no medical aid had been sought. He also knew that Rhymes and his wife Hannah had lived together unhappily; she was rumoured to be carrying on with another man, called John Grace. Breach also found out that Hannah had acquired arsenic by secretive means just after he had seen Rhymes. It was suspicious, and he told Hannah that there would be a coroner's inquest. She became agitated and said, 'What do people expect to find? I am sure he came by a natural death. However *I* gave him nothing.'

The County Constabularies Act had been passed in 1839. Rather than be policed by parish constables, who balanced their occasional duties with other jobs, for the first time rural areas could be policed by a professional force. But Berkshire County Constabulary would not exist until 1856, so it was down to Breach and East Hagbourne's non-professional parish constable to investigate.

Breach performed the post-mortem, and on discovering inflammation around the stomach and finding white powder adhering to its insides, he decided to carry out a test for arsenic. With some medical friends as witnesses, Breach performed two of the three precipitate tests, using ammoniacal silver nitrate and ammoniacal copper sulphate: both tested positive for arsenic. But he wanted a second opinion, so Breach and the stomach of Cornelius Rhymes took the train to London.

Taylor would later write a thirty-page article for *Guy's Hospital Reports* about the case. He began, 'Cases of poisoning by arsenic are, unfortunately, so frequent in the present day.' About half of it is taken up with thorough descriptions of the many tests they carried out. They tested the stomach contents first, and then they examined the stomach and found that something resembling arsenic had embedded itself in its lining. Taylor observed that arsenic causes the stomach to produce mucus in large amounts – presumably the irritant action of the poison forcing the stomach to defend itself. The result is that the stomach contents are vomited, but the arsenic becomes embedded in the mucus membrane and is very difficult to extract.

Some of their tests were carried out to second-guess the questions they might be asked in court. It was known that Rhymes had eaten gruel, so Taylor tested some of the stomach contents with iodine water. There was no reaction, so the gruel had left his system. Not surprising, Taylor remarked, if the deceased had vomited violently for ten hours before death.

Breach went to the coroner's inquest with tubes from Taylor's laboratory containing the arsenic that they had extracted. Hannah Rhymes and her paramour were sent to the next Assizes at Reading.

John Grace, who had been indicted as an accessory before the fact, was discharged when the grand jury decided there wasn't enough evidence for him to stand trial. Hannah had no such luck, and on 27 February, her trial opened. The prosecution set out their case against her, but were not allowed to refer to her 'improper intimacy' with Grace. Witnesses were called to attest to Rhymes's health, and Breach gave evidence about the path of Rhymes's illness.

The defence, in cross-examination, tried to suggest Rhymes died from diarrhoea, but once it became clear that the evidence of arsenic poisoning was incontrovertible, they had to explain its presence in Rhymes's remains. So they blamed Breach, insinuating that he had accidentally poisoned his own patient. Such things were not unheard of, but Breach explained that several patients had

had the same medicine without ill effect. He kept his arsenic away from his other medicines, and it was in solution: it was not in a powdered form that could have become muddled up in the powders that he prescribed for Rhymes.

The defence then challenged him over his youth. He had only been in practice for three years, and had he come across a case like this before? No, he had not. But his lack of experience didn't matter, as Breach had proof. He 'produced the results of the respective chymical tests, in small glass tubes, in which the metal of arsenic was clearly visible.' This was one of the Marsh test's great innovations, of particular use for the expert witness in court where the arsenic could be shown to the jury, not merely described.

Taylor, almost unknown to the press at this point and described as 'lecturer on chymistry at Guy's Hospital', corroborated the evidence of his former student. He explained that he and Breach had performed the Marsh test 'and two other tests, the most satisfactory that were known to the profession' and 'arsenic was reproduced by each of them.' Performing several tests was nearly their undoing; they had wanted to make sure there was little doubt of the presence of arsenic, whereas the defence asked why, if the reduction test was satisfactory, they had needed to test again. Taylor replied that 'the various tests were employed, not merely to corroborate each other, but for the acquisition of experience, on their general efficacy and general value.' Taylor had experimented on Rhymes's remains for the good of scientific knowledge.

Following Taylor came the comparatively prosaic evidence of the parish constable. He reported that on arresting Hannah, she had said, 'her husband had behaved ill enough to her while he was living, and it appeared he was going to behave worse now.'

And then the judge stopped proceedings. He was Sir John Taylor Coleridge, a nephew of the poet Samuel Taylor Coleridge. He'd heard the case for the prosecution and he believed that it was nothing but suspicion. He said that there was no motive, no proof that Hannah had administered the poison, and the whole case was that Rhymes was ill before the arsenic came into the house. The jury hesitated, but Coleridge went on: even if they found her guilty, he could not carry into execution the sentence of the law, so he urged them to acquit her. And so Hannah Rhymes was found not guilty.

Thanks to the intercession of Mr Justice Coleridge, a few months later Hannah appeared on the census living in John Grace's household as a servant. They were married in 1845, some distance from East Hagbourne; the cloud of suspicion had perhaps lingered.

Taylor justified writing about the case because there was a dearth of reports of British poison cases, unlike German and French cases in Continental journals. It was only Christison who was writing on the subject in Britain, Taylor claimed, and the Edinburgh toxicologist appeared often in the article. Taylor revered Christison's case reports: 'a medical practitioner who is called upon to give evidence in a Court of Law cannot do better than refer to them.' Taylor implied that there was room for another man beside Christison, and might he not be one Alfred Swaine Taylor?

The article was intended to be not just laboratory instructions, but a courtroom guide. Barristers displayed 'great ingenuity' in their cross-examination of medical witnesses. 'Each trial affords a lesson: it at least teaches us what we are expected to know, and what we must be prepared to answer.' He was providing his fellow medical men with the tools needed to survive their own day in court.

Taylor commented on some of the evidence in the case, even though, he admitted, it wasn't strictly medical. There was the issue of the vanishing arsenic. Hannah's nephew testified to having brought it to her house, but the next day, the arsenic had gone. Hannah said she had thrown it on the fire, and one witness said they had smelt it burning, and that it had been unpleasant, but another witness, who had smelt arsenic burn before, said that whatever Hannah had been burning, it wasn't arsenic. Taylor felt that the proceedings were rushed and that this part of the evidence could usefully have been gone into. The smell interested him from a chemical point of view, but he also felt that Hannah's explanation was nonsensical. Why go to the trouble of secretly buying arsenic to kill rodents, only to destroy it the next day by throwing it on the fire? Surely, Taylor remarked, this was an unconvincing fib, designed to conceal the fact that the arsenic had been used for nefarious purposes.

Taylor objected to Coleridge's claim that Hannah had no motive to murder Rhymes; she clearly did, and it was a problem that it was kept out of the evidence. Yet, Taylor wrote, a motive was not necessary for a conviction; the motive might be concealed, and murder can happen without any clear motive. Taylor said it was wrong of the judge to say that Rhymes had died of the diarrhoea that had begun before the arsenic came into the house. However, Coleridge perhaps felt that although it was proved that arsenic had been found inside Rhymes, it wasn't proved that it had killed him. Taylor disagreed; in his experience, diarrhoea could not kill a man so quickly if he was in good health.

Coleridge said that there was no proof of the arsenic being administered by Hannah Rhymes, but Taylor remarked that there were very few cases where there was direct evidence of administration. 'Poisoning is a crime almost always perpetrated in secrecy: the murderer, in general, so lays his plans, that there may be no witness to the act of administration.' If all cases of murder by poison required such proof, few criminals would be convicted.

Taylor could not think of another trial 'where the circumstances were so strong against a prisoner, and an acquittal followed'. He believed that it was wise for English law to admit circumstantial evidence: 'if it did not, it would allow the artful to escape, and punish only the ignorant and unskilled perpetrator.' Coleridge had declared that the case against Hannah Rhymes was nothing but suspicion, yet, Taylor asked, 'are we prepared to say, conscientiously, that there was nothing more than suspicion in this case?'

It is clear from the article that Taylor was building a network of scientists around him. Blood and renal expert George Owen Rees, a colleague of Taylor's from Guy's, appears in a footnote. He had read over the article before it went to press and, as a chemist, had highlighted a possible flaw in one of the tests. Rees had performed tests on sulphuric acid, which Taylor was using, and had shown that it was often contaminated with arsenic and selenium. That Taylor was happy to include this comment from Rees demonstrates Taylor's respect for him. Dr Thomas Grace Geoghegan, Professor of Medical Jurisprudence at the College of Surgeons in Ireland, had written that medical jurists should be made acquainted with the moral facts of a case before trial, because there might be insights that a medical expert could provide. When Taylor quotes Geoghegan, he refers to him as 'my friend'.

Taylor realised the Marsh test needed to be used carefully. He was critical of his former tutor, the famous Orfila, whose chemical evidence led to the conviction of Marie-Fortunée Lafarge in France in 1840. She was an unhappy woman in her early twenties, misled into marrying a man who claimed to be wealthier than he was. She had a motive, and when her husband died suddenly, she was under suspicion for his murder. Orfila was obsessed with the Marsh test, so when he was called in as the expert witness, he dissolved down the entirety of the dead man's body in acid so that he could perform the Marsh test on the resulting human soup. He did find arsenic – 1/130th of a grain – and Madame Lafarge was found guilty. Taylor didn't trust the evidence, writing that 'the employment of large quantities of nitric acid and nitre in an iron vessel, in stewing down the whole body of the deceased, might have accounted for the minute fractional quantity of arsenic deposited!'

Orfila's fondness for using the Marsh test to find tiny amounts of arsenic from huge samples led him to claim that human bone naturally contains arsenic. The minute quantity of metal that he produced from the test was too small for any of the confirmatory tests, so it could have been any metal, not necessarily arsenic. Taylor cautioned that Orfila had proved nothing, but had made a huge assumption. It was a concern, as it meant that defence teams in court could claim that arsenic found in murder victims was naturally occurring, based on Orfila's fallacy.

Taylor had given the Marsh test a run in an English courtroom, but a new test for arsenic was just around the corner.

Chapter 5

One of the Most Eminent Men
1842–45

In 1842, a confused correspondent wrote to *The Lancet*. They had bought volume one of Taylor's *Elements of Medical Jurisprudence* six years earlier, which promised a second volume. 'I have looked, year after year, for the fulfilment of the author's promise,' but recently the dismayed correspondent had seen an advert for a new title, Taylor's *Manual of Medical Jurisprudence*. 'Now, I think it right thus publicly to ask Mr A. Taylor, whether he intends publishing the toxicology &c., in a separate volume, to complete his former work, or whether he leaves the possessors of the first volume of his *Elements* to repurchase its contents.'

A reply issued forth from one John Churchill. Initially destined for the medical profession, Churchill had gained a taste for medical bookselling; working in a shop near St Thomas's and Guy's, he met Thomas Wakley and staff from the hospitals. He moved into medical publishing, keeping an eye out for potential authors, and he later published *Guy's Hospital Reports*. He knew that medical students needed textbooks, so he brought out a series of cheap but well-written manuals. Churchill approached Taylor and proposed that he write a book for him.

In his letter, Churchill explained that Taylor had 'made it one of the conditions that possessors of his *Elements* should be supplied with his new work at a considerable reduction from its published price, which will be lower than the price of the second volume, had it appeared uniform with the first.' Churchill said it would be like another of his titles, Erasmus Wilson's popular *Anatomist's Vade Mecum*. Conveniently, one of Churchill's books in this series was being published that very week, so he took the opportunity for a free advert by mentioning 'Mr Fergusson's *Surgery*'.

A new test for arsenic

German chemist Hugo Reinsch discovered a new test for arsenic in 1841. Two years later, having carried out several experiments, Taylor and Christison separately wrote about it for British readers. Taylor acknowledged that, while Marsh had made an 'ingenious discovery', it was not foolproof. It led to errors, as the amount of metal detectable was too small to be verified by other tests, which had led to

Orfila's claim that arsenic is found in human bone. Zinc's purity was unreliable; even though it could be tested, there was a risk that it would throw doubt on results, which clever barristers might take advantage of.

Marsh's test had been praised for its simplicity, but Reinsch's was even simpler; so much so, Taylor expressed amazement that no one had discovered it before. 'Reinsch remarked that a slip of bright metallic copper which he had placed in common muriatic [hydrochloric] acid became in a short time coated with a metallic film of an iron-grey colour.' He examined the acid using hydrogen sulphide and realised that the acid contained arsenic: this discovery 'led to the idea that copper and muriatic acid might be made a means of separating arsenic in the metallic state from liquids.' The toxicologist tasked with discovering whether or not a liquid contained arsenic could add hydrochloric acid to it, heat it, then slip in a piece of copper; any arsenic present would adhere to the copper. All of the arsenic could be removed by continuing to add copper until it no longer became covered in metallic film. The process was so simple that Taylor didn't even need to draw a diagram of the apparatus.

There were objections to the test: other metals also clung to copper when agitated by hydrochloric acid, but Reinsch explained how these could be told apart. Taylor wasn't as confident as Reinsch, particularly with antimony. Reinsch said it was 'a rich violet colour', and that arsenic only presented that shade when it existed in a small amount. Although antimony was lethal in large quantities, it was present in medicines given to treat gastric disorders that often had the same symptoms as arsenic poisoning, so it could cause confusion. Bismuth presented the same problem. Mercury and silver could also be detected using the test, but they were visibly different from arsenic and antimony.

Taylor warned that it was hazardous to pronounce on arsenic's presence based on appearance, but the 'extreme value' of Reinsch's process was that arsenic could easily be proved by heating the copper with its metallic residue in a reduction tube. If a white crystalline deposit formed, the ammoniacal silver nitrate and hydrogen sulphide tests could be used; no other volatile poison, Taylor said, crystallised in the same form. The certainty of the test would only fail when the deposit on the copper 'cannot be confirmed by other experiments applied to them'. Turning that metallic film into white crystals was vital 'to afford what the medical jurist requires, judicial certainty'.

There was a risk that hydrochloric acid might contain arsenic, but the test was perfectly easy: all they need do was add some copper to the acid before testing the suspect liquid, and they would know whether or not the acid was pure. Copper, Taylor confidently asserted, was perfectly pure of arsenic.

Taylor didn't have the opportunity to use the Reinsch test on a criminal case before writing up his paper, but what he did test it on gives us an insight into his laboratory. With no suicides on hand helpfully confessing to having taken arsenic, Taylor made use of a stock of stomach contents that he had stored from previous arsenic poisoning cases, some dating back nearly ten years. As disgusting as it sounds, Taylor's grim collection had come in handy. Besides, tests might have to be performed on exhumed bodies, so toxicologists had to be confident that the test could accurately detect poison in the contents of a stomach that had not belonged to a live human being for quite some time.

These weren't the only arsenic-tainted samples in his laboratory. There was 'a cake containing arsenic, which had been used in an attempt to poison', as well as the bottle of port wine from the dinner party. They were valuable because at a suspected poisoning, a medical witness might be asked to check for the presence of arsenic in food and drink.

Taylor wrote his paper in the July; Christison's followed in the September. He had used Reinsch's test in two criminal cases, 'where the bodies had been buried for four months, and I consider that it must soon supersede the beautiful but much more elaborate method of Marsh.'

His celebrity as a chemist

Reinsch had discovered an easy, quick test for arsenic and other metallic poisons. But Taylor's job was not an easy one when organic, or 'vegetable', poisons were used. In March 1843, due to his now well-known expertise, Taylor was asked to investigate the death of 5-year-old Mary Anne Vaughan, and in the December, the death of 1-year-old Thomas Samuel Ford; both had been given medicines containing opium.

When the inquest opened into Mary Anne's death, there was anxiety as the reputation of a medical man was at stake. It was suspected that a physician, Dr Johnson, had incorrectly prescribed for the child, who had been suffering from whooping cough. He hadn't visited the child, but this wasn't unusual practice. In order for Taylor to investigate, the inquest would be delayed for five days.

'That is a great length of time to allow a professional gentleman's character to suffer from groundless reports,' said a surgeon colleague of Dr Johnson.

A physician friend of Dr Johnson's said that it could surely take Taylor no more than six hours to complete his chemical tests. There were objections that the 'accusing party' were selecting the chemical analyst, but it was common at the time for the expert medical witness to essentially be on the side of the prosecution. The surgeon who performed the post-mortem tried to calm the situation by saying that he wasn't accusing anyone, and 'only selected Mr Taylor from his celebrity as a chemist'. Johnson and his friends 'admitted Mr Taylor's high standing as a chemist, but were only anxious that even-handed justice should be done.' Some inquest jurors chimed in that they believed chemical experts should be chosen by the other gentlemen too, and the coroner agreed.

When the inquest resumed, 'a great many medical and other witnesses were examined.' No opium had been found in the child's stomach, but some was found in the medicine bottle. The medical witnesses disagreed about how much opium could cause death. Taylor looked to his notes, and presented the inquest with a case from June 1832, when 'a child four and a half years old was killed by taking four grains of Dover's powders which were less than half a grain of opium.' He was asked by a juror if opium was the cause of death, but Taylor wouldn't swear to it. Looking at the contents of the bottle and how much was left, he calculated the amount of opium the child had taken. One of the physicians said the child had a heart condition, and that this, along with the opium and 'a severe cold' had caused her death.

Another medical witness explained at length that the dose taken by Mary Ann would possibly not affect a child in normal health. He suggested that the child should have been visited by the physician after she had taken the first dose of medicine and her condition deteriorated. But 'it was one of the misfortunes of "counter practice". The poor must be attended to.' Dr Johnson defended himself by saying he'd prescribed the same medicine many times, with no ill effect. The coroner said her mother was equally to blame for not seeking medical help after the first dose of the medicine had negatively affected the child. The jury's verdict was returned: Mary Anne's death was a tragic accident, caused by the mother administering too much opium, and Dr Johnson not properly instructing the parents.

In December that year, Mr Duke, the parish surgeon, was in the infirmary at Lambeth workhouse one morning when he heard a man snoring. But on investigation, he discovered 1-year-old Thomas, his stertorous breathing indicating that he was on the verge of death. Thomas lingered a few hours and died. Duke found an opium bottle nearby, so he performed a post-mortem, and passed the stomach contents to Taylor.

When the inquest into Thomas's death was opened, Duke was asked why he had involved Taylor. Apparently, the expense of hiring him was justified when a medical professional's reputation was at stake, but here was a pauper child who had died in a workhouse – was it worth the parish paying Taylor's fees? Duke said he thought the investigation would be of 'such paramount importance that he thought the best evidence should be given' and he was too busy himself. Besides, 'he had consulted one of the most eminent men in the profession, who had made the study of poisons his particular practice.' A child could die from a tiny amount of poison, so 'the analyzation must be most minute.'

The inquest resumed. Taylor had been given a bottle to examine and found that it contained six grains of opium; a quarter of a grain, Taylor explained, would be enough to kill a 1-year-old. He had not found any trace of opium in the child's viscera, however, but he hadn't expected to, 'because it would be absorbed almost immediately after it was administered, and digestion would be going on while the child lived.' He was asked by the coroner if opium left inflammation, and Taylor replied that it wouldn't; but metallic poisons like arsenic usually did.

The proof that Thomas had died from an overdose of opium had to come from the discovery of the bottle, his symptoms, and eyewitness testimony from paupers in the workhouse infirmary. They had seen three nurses complain about Thomas's restlessness at night, so they had dosed him up to make him sleep. The nurses had lied, saying that he had eaten a hearty breakfast, whereas the post-mortem showed he'd eaten very little, and the eyewitnesses said the child had stayed asleep. Some mercy was shown the nurses: they had not intended to kill him, so they would not be sent to trial. But their carelessness had cost a human life, and all three were sacked.

In both cases, Taylor had to admit that finding traces of organic poisons was extremely difficult, if not impossible. A child could die from a tiny amount of opium, which would be absorbed by the body; search as one might in the viscera, there would be no opium to be found. Metallic poisons could be absorbed by the body and not appear in the stomach, but even then, they could still be traced in the kidneys or liver or even in the brain. Organic poisons were not so simple.

By 1843, Taylor was a respected expert, known for his lectures at Guy's and the articles he had written, as well as one volume of his uncompleted, but valued, book. As the decade wore on, his fame would increase.

We warmly recommend Mr Taylor's treatise

Taylor's book, which Churchill had promised two years earlier, was eventually published in early 1844. Called *A Manual of Medical Jurisprudence*, it was well received by *The Lancet*. It was a condensed version of a longer work that Taylor had prepared for publication, the updated first volume of *Elements of Medical Jurisprudence*, with its long-awaited second volume.

The Lancet reminded readers of the history of medical jurisprudence teaching in Britain. They quietly praised the Apothecaries' Company, when ten years before they lambasted them as 'the Hags of Rhubarb Hall'. Books on the subject had been published in France 'of greater or less value', and although several treatises and essays had been published in Britain, 'there has been no work which has succeeded in becoming a standard authority on medico-legal subjects with the exception of Professor Christison's celebrated *Treatise on Poisons*.' For English language texts, the British were having 'to apply to our transatlantic brethren, and to adopt, as it were, Dr Beck's treatise'.

Taylor 'is favourably known to the profession by previous publications on forensic subjects,' and throughout the book 'we find evidence of a clear, vigorous intellect, able to separate the true from the false, the probable from the improbable, and able to contribute, by his personal views and researches, to the progress of the science on which he writes.'

Poisoning was, unsurprisingly, 'treated at considerable length, and forms a very elaborate and valuable digest of the present state of science'. *The Lancet* picked out the chapter on arsenic for particular praise. The *Manual* covered many more subjects than poisoning, however – wounds, drowning, infanticide, even lightning strikes – which meant that Taylor had produced the most comprehensive British text on medical jurisprudence up to that point.

Taylor's *Manual* was illustrated with recent cases, which impressed *The Lancet* so much that it inspired them 'to give our columns a monthly synopsis of all the important medico-legal data, &c., which come to light'. There were facts in Taylor's cases that 'appeared to many of trivial importance' but put together formed a valuable, impartial record.

In order to pack so much information in, Churchill used very small type. *The Lancet* remarked that this was 'very close but beautifully clear'. However, the *Provincial Medical and Surgical Journal* (the forerunner of the *British Medical Journal*) complained in the arch, sarcastic tones typical of the period: 'We must confess that, when we glanced at the closely printed page and microscopical type, the very thought of attempting the perusal made our eyes ache.'

Having carped about the typesetting, the reviewer praised Taylor for explaining the dangers to children's lives posed by the 'empirical and incautious manner in which [opium] is so commonly exhibited by ignorant nurses, or careless parents and relations'. This was the main thrust of their review, ignoring Taylor's innovative approach to medical jurisprudence. But all editors and reviewers have their axes to grind, and in Taylor's section on opium, their reviewer had found an opportunity to give vent to their views.

The *London Medical Gazette*, meanwhile, was fulsome in its praise. Taylor's *Manual* was, they said, 'one of the very best books of any description we had ever read'. An author can hardly receive a better review than that.

Taylor's approach, collating cases from Britain and across the world, required an organised system. He kept diaries and commonplace books (a sort of scrapbook), which he catalogued and indexed, so that he could arrange new information, ready to update forthcoming editions. He was a pioneer in a new field, and over his long career, he collected over 400 medico-legal tracts and books, in English, French, and German, which he drew on for his books and articles. He didn't necessarily agree with all the writers, as acerbic marginalia attest: 'absurd', 'sophistry', and 'the profundity of ignorance'. There are signs that he checked the data used by the other writers, adding his own notes. He bound the texts into twenty-nine volumes, each with a table of contents, so that he could easily pilot his way around his vast library.

While the public Taylor appeared in the witness box and the lecture theatre, and could be found in his laboratory at Guy's, in entirely all-male environments, the production of his books was partly the work of the private Taylor, at home in Cambridge Place. His wife, Caroline, helped him to revise his work for publication. This, at a time when women could not train to become doctors, could not become barristers or judges, and were not called for jury service. Taylor's *Manual*, which was a first in British publishing, was partly the work of a person whose sex prevented them from having a public role in the very sphere that the book treated: unless she found herself witness, victim or accused.

Does arsenic float on tea?

A month after *The Lancet*'s review, the Taylors had another reason to celebrate: the birth of their daughter, Edith Caroline, on 21 May 1844. She was baptised at St Pancras church, as her brother had been eight years earlier.

A week after Edith's birth, Margaret Leaver lay ill in her bed on Bermondsey Street, near Guy's Hospital. Her son George brought her a cup of tea, made by John, her husband. It was almost too hot to drink, and too sweet for her palate. This being the sole foundation of her suspicion, she took the cup to Guy's Hospital, and William Tiffen Ilet, assistant to surgeon Bransby Cooper, poured the liquid out of the cup and found a sediment at the bottom, 'three or four grains of finely powdered arsenic'. He gave Margaret an emetic and examined what she brought up, but could not find any arsenic, and neither could he find any in the liquid of the tea.

At the Old Bailey, 11-year-old George pointed the finger of suspicion at his father. His mother's cup had been laid out on the table, and when George had returned from buying a penny loaf, a saucer had been placed on top of the cup, the implication being that it was hiding the arsenic. After putting tea in the cup, George saw his father 'skimming the top off the tea and throwing what he was skimming off on the carpet'. When John was arrested, he was told that he was being charged with attempting to administer poison to his wife, and John replied that 'he knew nothing of it, that it was a planned thing between his wife and her companions.'

In their evidence, George and Margaret said the police found arsenic in their shop's till without their prompting, whereas the police said Margaret had told them where to find it. The Leavers had rats in the house and shop, and were in the habit of leaving arsenic on bread in the rodents' holes to kill them. Soda, which in appearance could easily be muddled up with arsenic, was sometimes used by the Leavers to 'draw' their tea. It was enough to cast doubt on the case, and John Leaver was found not guilty.

Taylor did not give evidence at the trial, but Ilet must have consulted him, and asked if arsenic floated on tea. Was that what John had been skimming off onto the carpet? When referring to the case, Taylor later wrote, 'I have detected no arsenic dissolved in tea when it was abundant in sediment.' Boiling arsenic in distilled water would not make it dissolve; 'it partly floats in a sort of white film, and is partly aggregated in small white masses, at the bottom of the vessel. It requires long boiling, in order that it should become dissolved and equally diffused through water.' Arsenic, therefore, was difficult to disguise in tea.

John Leaver's liberty depended on the question, as the fact that the arsenic hadn't dissolved would suggest at what point the arsenic had been added: was it put in the kettle, or the pot, or into the cup? It could indicate, although not necessarily prove, whether the introduction of arsenic was an accident. But with arsenic being so easily procured, and food and drink often being vehicles for poison, the issue of arsenic floating on tea was an important one to resolve.

Unprincipled '*Principles*'

In October 1844, Dr William Augustus Guy, professor of forensic medicine at King's College London, and assistant physician at King's College Hospital, wrote to *The Lancet*: he was defending his reputation.

The first part of his work *Principles of Forensic Medicine* had been published at the end of 1843, with later parts being published in 1844. Readers who had read Taylor's *Elements* and more recent *Manual* were surprised to note similarities between them. Guy claimed it was an 'inadvertence', and that the passages had come from Taylor's earlier *Elements*, because 'in preparing my first course of lectures, I made extensive use of Mr Taylor's work, and embodied several passages without taking the precaution of noting the source from which they had been derived.' A friend had already pointed out the error to Guy, but the mistake 'has since become one of more than private remark', so Guy felt it prudent to come forward, even though it was extremely embarrassing. Taylor had temporarily filled the vacant seat at King's, until Guy's appointment. Guy's letter states that he is 'under personal obligations for [Taylor's] friendly assistance rendered to me at my first entrance on my duties at King's College'.

In November, the *London Medical Gazette* reviewed Guy's work, at once picking up on the similarities it bore to Taylor's. 'We became strangely possessed with the idea that we had already read the book; every illustration seemed as familiar to us as the face of an old friend.' Guy tried to explain himself in their pages; using courtroom language, he pleaded 'not guilty'.

Taylor expressed his sadness at what had happened, and was clearly hurt that Guy was trying to evade the accusations laid against him. Guy 'has put in his plea of "not guilty", and the verdict, according to all the rules of justice, must rest with others, and not with himself.'

The Lancet was unrestrained when they reviewed Guy's book a few months later in February 1845. They declared 'that a grosser piece of plagiarism, or a more flagrant breach of literary honesty, has never come under our observation'. In his *Manual*, Taylor had set out 'rules to be observed in investigating a case of poisoning'. This was an entirely new scheme, yet Guy's book included the near-identical 'rules of medico-legal examination in cases of poisoning'. *The Lancet* compared extracts side by side, stating that 'the reader will at once perceive that Dr Guy's section is nothing more than an epitome of Mr Taylor's chapter, with the materials variously transposed.' They were forensic in their comparison, finding point after point that they felt indicated plagiarism.

Guy's explanation of having made an error in not recording the sources for his lecture notes earned him the retort, 'What miserable shuffling is this! ... Sincerely do we hope that we shall never again encounter such unprincipled *Principles* as those of Dr Guy.'

The situation was not severe enough to scupper Guy's career: in 1846 he was made dean of King's College medical faculty, and he went on to edit several journals. He was appointed to various lectureships and royal commissions, and was vice-president of the Royal Society for two years. His *Principles of Forensic Medicine* went into several editions and became a standard work. But the furore showed that Taylor had vociferous allies.

Dear Mr Editor ...

Taylor's reputation was such that the following year, he became the editor of the weekly *London Medical Gazette*, a post he would hold until 1851. It gave him a platform from which to hold forth: his personal foibles were never far from the readers' notice.

Under Taylor's editorship, the journal consisted of lectures given by medical professionals, including Taylor's own. Medical men submitted interesting cases, including diagrams of organs and body parts, and wrote in to comment on their experiences in general practice and in hospitals, or as witnesses at inquests. These were interspersed with book reviews, Taylor's collection of tracts and books growing exponentially during his editorship as he was personally sent titles to review. His write-up of a book about water cures at Malvern was scathing, 'a work on the fashionable quackery of the day'. When a fifty-six-page pamphlet came his way, which was essentially a demolition of Christison and Thomson, Taylor used his review of it to defend his Edinburgh peers.

There were snippets that show the breadth of Taylor's reading – a parliamentary roll from 1422, 'an ancient relic of the suppression of quackery', and an extract from a 'caution to druggists – the duties of the ancient poticarye' from 1562, both included to show readers that the fight to regulate medicine was a long one. One surgeon wrote in about a 4-year-old boy who survived after accidentally swallowing a musket ball during a playground tussle, and there was a case Taylor had found in a French medical journal where a family had eaten turnips that had strange hallucinogenic properties. Were they turnips at all, Taylor wondered. There was news from Prussia that sales of powerful medicines were to be restricted owing to accidents caused by mistakes in prescriptions. 'Contrast this with the free trade in the sale of poisons which is openly carried on in Great Britain!' Taylor exclaimed.

There were inquest and trial transcripts, sometimes for medical malpractice, but often for murder cases, such as a trial in Edinburgh for arsenic poisoning that resulted in the Scottish verdict 'not proven'. Taylor argued that the case showed how difficult it was to establish murder by arsenic due to the lack of direct proof, which 'is contrary to the very nature of the crime'. It also demonstrated, he said, 'the want of a check upon the sale of poisons'. The accused woman had bought 'a pennyworth of poison, enough … to kill four persons' from 'a *Doctor of Medicine*'. Taylor's horror is evident in his italics. It was known that juries would shy away from a guilty verdict if they felt strongly against capital punishment, but Taylor suggested that medical jurists might be skewing evidence to avoid guilty verdicts as well. This wasn't going to happen on Taylor's watch: 'We shall take care to expose any case of this kind that may come before us,' he thundered.

James Marsh was remembered in an obituary. Although he 'was not a member of the medical profession, we think it only a fair tribute of respect to his memory to insert this notice in our pages.' He had, after all, found 'an ingenious application of chemistry to toxicology', so Taylor had a lot to thank him for.

Henry Letheby submitted pieces to the *Gazette*, even though he and Taylor did not always see eye to eye. In debates over the use of ether, Dr John Snow appeared; he went on to administer chloroform as pain relief for Queen Victoria during childbirth. Taylor's friend John George French contributed an article about cholera, and George Owen Rees, Taylor's colleague from Guy's Hospital, wrote a paper on Bright's disease.

Taylor's editorials reflected the contemporary concerns of his professional peers. There was the medical reform bill; Wakley was making a case to Parliament for the public registration of anyone calling themselves a surgeon, physician or apothecary. 'The hearty desire of the great body of English practitioners is, we verily believe, to clear the profession of all these quacks and impostors.' There was discussion on quarantine laws and how they could balance public health and commerce; as a medical man from a family of merchants, Taylor understood the concerns of both. He commented on the alarming increase in diarrhoea in London over the summer of 1846, which the press had claimed to be the return of Asiatic cholera, thus leading to panic. Taylor argued, from investigations carried out by medical experts, that it wasn't cholera at all, and was merely an increase in fatal stomach upsets caused by an uncharacteristically warm summer. By the winter, deaths had increased from bronchitis and other chest problems, caused, Taylor said, by the unusually cold season.

Taylor had much to say on the Health of Towns Bill, specifically on the issue of burials in London. 'There is no graveyard in London which is not closely surrounded by dwellings,' and all the noxious gases from decomposition were going into Londoners' homes. There had been scandal when churchyards were being cleared to make way for internments, dustmen carting off parts of coffins and former residents. How, Taylor demanded, could new sewers, necessary for the health of the living, be tunnelled under London, when they had to be dug through the resting places of the dead?

Taylor commented on the letters and articles submitted by his readers, who seemed impressed by the results of using ether for pain relief. But when it turned out that attempts were being made to patent the process, Taylor's ire was stoked. He mentioned Daguerre, who still annoyed Taylor-the-photographer because he had dared to patent 'the use of solar light, rare as it is, in England!' If using ether was patented, Taylor angrily pointed out, it would prevent the poor from having access to pain relief. His tone was heavily sardonic: anyone who patented it 'can look for a satisfactory return only to legs and arms of the wealthy part of the community'.

Coroners' inquests were a frequent topic of his editorials. The system for appointing coroners was flawed, Taylor argued, as it was done via election, and he strongly believed that a coroner had to be a medical or legal professional. A statute had been passed that meant coroners had to pay for the expenses of an inquest first, before submitting their accounts to the magistrates at the quarter sessions. Taylor declared it a 'vicious principle of making the qualification of a coroner to depend upon *property*, and not upon *skill* or *experience* in the intricate questions affecting life and liberty'. It shut out the role from anyone not independently wealthy.

The cost of inquests often came up for debate in newspapers, local ratepayers objecting to the expense. Coroners cut corners; as medical witnesses had to be paid for attending inquests, and police officers were not, often the medical attendant who witnessed someone's illness and last moments would not be called, even though they could offer important evidence. There were many times when post-mortems hadn't been performed, but should have been. Taylor believed that all coroners 'should produce proof that they have studied medicine, or at least that branch of it known as forensic medicine'. He wrote that tightening up inquest procedures would 'prevent those disgraceful struggles for fees which now exist – lead to more certain detection of crime – obviate the necessity for frequent disinterment of the dead'. His recourse to dashes emphasised his strength of feeling.

Taylor's intense dislike of barristers, founded on his own experience in court, often had an airing in his editorials. He gave a recent example where a barrister had rudely asked a witness, 'Come Mr Medical-man, do you mean to answer that?' Taylor claimed that barristers were able 'to use language, which it might be libellous in a printer to circulate, although he might give a verbatim report of proceedings'. The barristers were 'protected by a wig and a gown' while the medical witnesses were publicly anatomised. Taylor felt that 'calumny and slander' being used as 'weapons necessary for the defence of a client' prevented 'good men' from appearing as witnesses. The unchecked behaviour of barristers, who would not be reined in by judges – after all, they had once been barristers themselves, Taylor said – was 'a most injurious influence on the course of justice'.

Despite the Anatomy Act of 1832, it was becoming difficult again for medical schools to get hold of cadavers, partly because more medical schools had opened. It was a pressing problem as practical anatomy was 'the only sure basis of medicine and surgery'. Taylor felt that too much power remained with the parochial authorities who supplied the cadavers, and that 'ignorance, vulgar prejudice against dissection, and, what is worse, venality [interferes] with a regular supply of subjects to our dissecting rooms.'

Under Taylor's editorship, the *London Medical Gazette* reflected the interests of its readers, but it is also an intriguing display of Taylor's own hobbyhorses. Inquests, trials, anatomy, the temerity of anyone to patent a process that was good for the world, the strange and the curious, poisons and poisonings, all flowed from his often irate pen.

Chapter 6

My Heart is as Hard as a Stone
1845

The occasional leisure of four years

When not making photogenic drawings, or editing a medical journal, what was a professor of chemistry and medical jurisprudence to do with his time? In 1845, Taylor drew up an exhaustive comparative temperature scale for Fahrenheit, Centigrade and popular European scale Réaumur. The paper thermometer had 'been designed to obviate the necessity for those perplexing calculations' required to work out the different methods used in Britain and on the Continent.

Taylor heavily annotated his scale, and explained it in a short pamphlet, to 'convey information on numerous interesting points, connected with temperature in relation to Climatology, Physical Geography, Chemistry and Physiology'. He thought it ridiculous that thermometers still marked old points such as 'Summer-heat, Blood-heat, Fever-heat, Spirits boil' as science had advanced. Instead, his thermometer told you the mean temperature of Stockholm, the temperature at which tallow melts, at which starch converts to sugar, of the Durham coal mines 900 feet down, of various bodies of water across the globe and at different depths, the temperature at which water boils at different elevations …

One review said, 'we think the profession very much indebted to Mr Taylor for this most useful table.' It had taken four years to compile, by consulting 'the best authorities', and that 'many of the facts he was enabled to collect or verify by personal observation, and some of the chemical phenomena have been derived from direct experiment.' At least it had kept him occupied in his laboratory until there was a suspicious death to investigate.

I can't confess what I've never done

On 25 January 1845, Taylor had a visitor at Guy's Hospital. Arthur Lamb was a surgeon, from Newbury in Berkshire. The day before, the body of 3-year-old Eleazar Jennings had been exhumed from the churchyard of Thatcham, near Newbury. He had died on Christmas Day 1844, from an illness that came on suddenly and took his life in the space of forty-eight hours. An inquest had been held two days later;

without a post-mortem, a verdict of natural death was returned. A fortnight later, his brother Henry died suddenly as well. Another inquest was held; this time, there was a post-mortem, and inflammation was noticed in Henry's stomach and intestines. A rough chemical analysis of his stomach contents was carried out, but no poison was found. A verdict of death by natural causes was returned and Henry was reinterred.

Lamb had assisted at the post-mortem. He had a strong suspicion that poison was present in Henry's case, so he sent Taylor a phial of Henry's stomach contents. Taylor analysed it and found a small amount of arsenic; the seasoned toxicologist was able to find what the provincial surgeons could not. His discovery led to Eleazar's exhumation. The boys' father, Thomas Jennings, was present in the churchyard and identified Eleazar's body. When someone remarked that there was suspicion that both children had been poisoned, Thomas said, 'there was no more poison in them than there was in him.'

Taylor and Lamb examined the child's viscera. There was a solid white substance in the stomach, which turned out to be suet. Along with the partially digested currants and raisins in his stomach, it was clear the tragic child had eaten a plum pudding; it was Christmastime, after all. There were signs of inflammation, and there was ulceration under some yellow matter, which pointed towards poisoning. They tested the stomach contents, using the Reinsch test, two precipitate tests and the hydrogen sulphide test; they all showed the presence of a tiny amount of arsenic. They looked at the yellow deposit under a magnifying glass and could see 'white-looking crystalline particles, like sand'. The yellow matter was arsenic; originally white arsenic, it had turned to arsenic trisulphide as the body decomposed.

From the unnatural redness of the stomach, and the ulceration beneath the yellow arsenic, it was clear that the poison had been taken by Eleazar while he was alive, and that it had killed him. Taylor observed that there were from five to six grains of arsenic in the stomach; this was what was left after Eleazar had suffered from vomiting for two days. 'I may remark,' Taylor wrote, 'that the quantity of arsenic which would probably suffice to kill a child of this age would be about two or three grains' – even less, in some circumstances.

Lamb thought he knew who had killed Eleazar. On 6 January, he had been in the Jennings's cottage, and had brought up the issue of arsenic. Thomas had told him that he had had no arsenic in the house for years, but that he did have some in his gamekeeper cottage in the wood. But it was eyewitness testimony from Thomas's own niece, Maria Carter, that would place him in the dock.

Maria was 13, and had lived with the Jennings for two years, helping to look after the young family. There were four Jennings children, and at the time Thomas's wife was pregnant. Two days before Christmas, Mrs Jennings had been out at dinnertime while the rest of the family had sat down to a meal of bacon and potatoes. There was a salt cellar on the table, and everyone had used it, except Eleazar. As they tucked in, Thomas went into the pantry, returning with a pinch of something white between finger and thumb; Maria remembered it looking like salt. Thomas put it on the edge of Eleazar's plate, and the boy dipped his potato in it and swallowed it whole.

About half an hour after dinner, Eleazar's illness began with pain in his belly, and the next day he was sick. He didn't appear to have had diarrhoea, although he did go 'once or twice to the necessary'. He was thirsty and drank mint tea. By Christmas Day, the vomiting was worse. Mrs Jennings wanted a medical man to see the child, and the father sent for one, but he did not arrive in time. Eleazar died about midday.

Now that there was medical and chemical evidence to prove that the child had died of arsenic poisoning, his father's actions took on a horrible significance. On 7 February, the jury of the resumed inquest found that the child had died from poison 'administered wilfully by his father Thomas Jennings, for the purpose of destroying life'. Thomas was sent to jail to await the next Assizes.

God's lightning

As shocking as the Jennings case was, it earned few column inches. It was up against a far more sensational poisoning: a well-to-do Quaker had murdered his mistress. If this wasn't enough to capture the attention of the Victorian public, it was a modern crime, involving an escape by train, an alert sent by telegraph, and a poison that had been little used up until that time for murder.

On New Year's Day 1845 – a week after Eleazar Jennings died – the peace and quiet of Salt Hill, Slough was rent by a scream. Sarah Hart's neighbour ran to check on her; the scream had come from Sarah's cottage. The neighbour encountered an agitated man in Quaker dress, hurrying away. Sarah was found insensible on the floor, her clothes in disarray, indicating that there had been a struggle. Local surgeon Henry Montague Champneys was summoned, but nothing could be done. Sarah's pulse faltered and stopped not long after he arrived. For reasons that Champneys never fully explained, he opened a vein in her arm.

The hunt was on for the fleeing Quaker; he was not difficult to track down in his distinctive clothing. Champneys' cousin traced him to the railway station, and saw him board a train to Paddington in London. He alerted the stationmaster, who used the telegraph to pass the description of the wanted man to the railway police. As there was no letter Q on the machine, the message contained the word 'KWAKER', which the station staff in Paddington scratched their heads over. The message had to be repeated, until they realised they meant 'Quaker'. The suspect was trailed from Paddington by police, and later arrested. His name was John Tawell.

The medical men were busy trying to work out what had killed the otherwise healthy Sarah Hart. Champneys, and another Slough surgeon, Edward Weston Norblad, performed a post-mortem, 'there being strong reasons for supposing, in the absence of any external marks of violence, that the deceased's death was occasioned from the effects of some potent poison.' Her viscera would have to be examined, as well as a phial found in a garden along the route taken by Tawell as he made his escape.

The surgeons headed to the London laboratory of John Thomas Cooper, a lecturer and a chemical manufacturer. Cooper was among over seventy other scientists, including Rees at Guy's Hospital, who had founded the Chemical Society. He had been consulted on legal cases before and was no stranger to the courtroom.

Sarah's stomach contents were acidic, so they started their investigation by performing tests for oxalic and sulphuric acids; they found neither. Nor did they find any metallic poisons. So they tested for prussic acid, in the stomach and in the phial from the garden: and they found it.

Prussic acid, or hydrogen cyanide, is a highly volatile compound of hydrogen, carbon and nitrogen. It was isolated from the pigment Prussian blue; in Germany, it was known as 'blue acid'. When sold in shops, it was in a solution with water or alcohol, so was available in varying strengths, and was used medicinally for a variety of illnesses. It had been used by suicides, twenty-seven deaths recorded from 1837 to 1838, but by the 1850s, Taylor could write, 'of late years it has, however, acquired a fatal celebrity as a means of murder!'

It was known that prussic acid smelt of bitter almonds, like marzipan or amaretto; bitter almonds themselves contain hydrogen cyanide. Research carried out in the 1950s and 1960s showed that not everyone can detect the smell; the ability is a genetic trait. At post-mortem, the smell could be obscured by strong odours such as tobacco or essential oils. This caused problems for toxicologists, who could conduct laboratory tests to prove the presence of prussic acid, but would have to contend with courtroom cross-examinations demanding to know whether they could smell bitter almonds, despite it being an unreliable method of detection.

At the inquest, Champneys said he didn't detect the smell, and justified this by saying that it must've been a salt of prussic acid. He claimed that he had opened Sarah's vein to see if he could smell it, and said he hadn't smelt anything at the post-mortem. Norblad similarly said he couldn't, but he too was adamant: Sarah Hart had been poisoned by prussic acid.

The police had shown their worth, tracing the druggist who had sold the phial found in the garden. The druggist was taken to Tawell's cell, where he positively identified him, as did Sarah Hart's neighbour. An eyewitness said that Tawell had been on an omnibus in Slough, and, perhaps trying to shake off a trail, asked for directions to Sir John Herschel's house. Given that the polymath Herschel had used a compound of cyanide in his cyanotypes, this was somewhat ironic.

The respectable Quaker proved to be nothing but when his previous life was reported in the press. He had converted to the Society of Friends but was ejected after marrying a woman who wasn't a Quaker. He had been convicted of forgery and transported to Australia, where he had run a druggist's shop; no wonder he selected poison as his weapon. He had returned to Britain with money. Sarah Hart had been his mistress since his wife's illness, and after his wife's death he had married a Quaker schoolteacher; since that marriage, he had fathered two children by Sarah. He was in financial difficulties and maintaining Sarah was a drain on his purse. Here, then, was a motive. Despite Tawell hiring a barrister and several solicitors to attend the inquest, the jury decided that he should stand trial for Sarah's murder, and he was remanded in custody until the Buckinghamshire Lent Assizes.

What a paradox does such conduct exhibit

The Berkshire Assizes in Reading opened first; Thomas Jennings stood in the dock on 5 March, accused of murdering his son. Unlike most poisoning cases, there was eyewitness testimony of the accused apparently administering the poison. This, combined with Taylor and Lamb's evidence, presented a strong case for the prosecution.

Taylor felt that the appearance of the stomach was extremely well marked, demonstrating unequivocally that the child had died of arsenic poisoning. Rather than rely on description, Taylor recruited the medical illustrators at Guy's to draw the appearance of the stomach, and to use yellow arsenic mixed with gum to show the court exactly what he had seen. Besides showing test tubes and copper gauzes to the jury, it was another way to explain how he reached his conclusions. The defence asked, wasn't arsenic a normal part of the human body, as demonstrated by the celebrated Orfila? But the judge, on whom the illustration of the stomach had made an impact, asked sarcastically, was it medically possible for that much arsenic to be spontaneously generated in the stomach? No, it wasn't.

Jennings argued that his niece was framing him, and implied that she had killed Eleazar. But the jury found him guilty, and Jennings would hang at Reading, from William Calcraft's noose, on 22 March. He denied his guilt right up to the end. As his wife left him for the last time, she begged him to tell her if he knew anything about the death of their child, and Jennings replied, 'I can't confess what I've never done.' Mrs Jennings cried, and her husband said to her, 'My heart is as hard as a stone; I wish it was as soft as yours!'

A week after Jennings's trial, Tawell appeared in the dock about 30 miles away in Aylesbury. The case had attracted much attention, particularly due to fascination with his double life. 'What a paradox does such conduct exhibit; and on what principles can it be explained?' one newspaper pondered.

The trial lasted for three days, *The Times* beginning each day's report carping about the poor accommodation for members of the press. A man of more means than Jennings, Tawell's defence team had subpoenaed chemical experts to appear, hoping to undermine the evidence given by the prosecution. Fitzroy Kelly was Tawell's barrister, who was only months away from being knighted and made Solicitor General. Beside him, as Kelly made his cross-examinations, was Henry Letheby. Like Taylor, he had attended medical school and had an interest in chemistry; he was also a lecturer in medical jurisprudence. In the wings were seven more subpoenaed experts, including William Herapath, a professor at Bristol Medical School, and two of Taylor's colleagues from Guy's.

Champneys, Norblad and Cooper had little if any experience of dealing with deaths by prussic acid, so they referred to textbooks, as did counsel for the prosecution and defence. Taylor may not have been present, but his book was. During Kelly's cross-examination of Cooper, he asked, 'Are you not reciting the case mentioned by Taylor, of a boy who had been killed by three and a half grains of prussic acid?' Letheby was hunting through a pile of books for quotes for Kelly to use.

Much emphasis was placed on prussic acid's distinctive smell. Perhaps this explains why the surgeons, who claimed at the inquest not to have detected the odour, now claimed that they had. Champneys even went as far as to say that he had opened Sarah's vein to sniff for prussic acid. Cooper pointed out that the smell could have been obscured by the Guinness stout that Sarah had drunk. But whether the smell was detected or not, the chemical tests had shown the presence of prussic acid.

Kelly, presumably prompted by Letheby, asked if the prussic acid could have come from the apple that Sarah Hart had been eating. There was a barrel of them in her house, and she had been eating one before her death. But the veteran Cooper had pre-empted this, by trying to extract prussic acid from apple pips. From fifteen apples, he had extracted a tiny amount of cyanide, certainly not enough to kill; therefore, the amount in one apple couldn't have killed Sarah Hart.

Cooper's scientific evidence was so solid that there was no point in summoning the seven other experts. The defence tried to claim that Sarah had committed suicide, but there was no evidence to support it, and Tawell was found guilty. He confessed to her murder before he was hanged by the ubiquitous Calcraft on 28 March.

Those short and easy roads to knowledge

Taylor conducted an experiment, involving 'one of the witnesses for the prosecution, and another for the defence' – presumably Cooper and one of his Guy's colleagues who had been subpoenaed. He put two-thirds of a grain of prussic acid, a quantity sufficient to kill an adult, in 3 ounces of porter, and asked them to smell it. Neither could detect the odour.

Taylor took to the pages of *Guy's Hospital Reports*. Even before the Tawell case had come to court, he had decided that prussic acid would be a good subject for one of his medical jurisprudence articles. His focus was a case he had worked on in March 1844, when Mr Newham, a former Guy's pupil, sent him a stomach and a phial to test. A travelling salesman called Thomas Hobson had apparently committed suicide at an inn in Bury St Edmund's, Suffolk, and it raised questions about prussic acid and the law. How quickly did it act? How much time could pass before tests proved useless?

Hobson had been found lying placidly in bed, the blankets neatly drawn up to his shoulders. The phial had been recorked, and placed on a chair behind the bed. It was assumed by some that prussic acid worked very quickly and induced convulsions and a shriek, but Hobson had apparently not had convulsions, no one had heard a shriek, and he'd had time to cork the bottle, reach over to place it on the chair, and pull up the blankets. It was an important consideration: suicide by prussic acid could be misconstrued as murder.

Taylor didn't receive the stomach and the phial until twelve days after Hobson's death. By then, the inquest, based on evidence other than Taylor's test tubes, had already found that it was a *felo de se* (suicide), and Hobson had been buried without a funeral service. Taylor decided to carry out his tests anyway, the circumstances of the death fascinating him.

The surgeon who carried out the post-mortem said that most of Hobson's body had smelt of prussic acid, but the stomach and its contents, by the time Taylor received them, had lost their odour. Neither had the poison shown up in his tests. Taylor wasn't surprised; prussic acid's volatility meant it hadn't been detected by tests as late as even eight days after death.

Taylor's well-researched article comes to the conclusion that the smell of prussic acid doesn't always accompany its presence, and that its presence isn't always accompanied by its smell. He added that he didn't think animal experiments, which toxicologists performed many of, were to be relied on too heavily when applying the knowledge to human beings, particularly in medico–legal inquiries. These experiments had led to the fallacy of the 'shriek' in prussic acid poisonings.

As the article was published at the beginning of April, he was able to add a short appendix on the Tawell trial. Taylor was unhappy that the defence counsel had 'attempted to prove that in death from this poison the odour must always be perceptible' and that they had tried to 'throw discredit upon certain cases recorded in most medico-legal works, the facts connected with which were against this presumption'. He believed this had partly arisen due to mistranslation, German and French texts having been quoted at the trial, sometimes where the original had been in German and then translated into French, and then into English in the courtroom.

A few days later, Letheby wrote to *The Lancet*. So much had been written of late about prussic acid, he said, for three reasons:

1st   To cover a hasty and an unsound opinion.
2ndly   To justify a careless quotation or an erroneous report.
Or, 3rdly   To gratify the *cacoethes scribendi*.

Letheby's sarcastic Latin was a pointed way of mocking Taylor's frequent appearances in print; it was the man's 'insatiable desire to write'.

Letheby may have felt undermined by Taylor performing his own experiment related to the Tawell trial, and he was unhappy that Taylor's latest piece in *Guy's Hospital Reports* had touched on two cases he had been involved in: Tawell, and the trial of James Cockburn Belaney. Belaney was a surgeon, accused of murdering his wife with prussic acid; at the trial, Letheby had been questioned on the 'shriek'.

Letheby complained that the surgeons at the Tawell trial had changed their testimony about whether they had been able to smell the prussic acid, and wondered what the devil Champneys was about when he cut open Sarah's arm. 'The whole case is one of blunder and misrepresentation,' Letheby thundered; which he would say, as the defence had failed.

There was too much reliance on 'certain works upon medical jurisprudence', and Letheby questioned whether or not these texts were accurate – a response to Taylor's assertion that in some cases they weren't. Letheby announced that it was Taylor himself who was guilty of mistranslations. Even if that were true, Letheby seems to be covering himself, rescuing his own reputation by attacking Taylor.

Early in April, *The Lancet* had published an editorial on the medical aspects of the Tawell trial; they found it lacking. One of the cases referred to at the trial was one of seven epilepsy patients in a Paris hospital, who had been accidentally overdosed with prussic acid. It appeared in many texts, but every writer had a different interpretation of it. 'A more perfect demonstration cannot be desired of the evil of trusting for information to compilations and manuals, those short and easy roads to knowledge.'

But what could provincial surgeons do when faced with such an usual case? For all that Champneys was mocked for his lack of knowledge, he was beside

Sarah as she died, and he had realised that it was unlikely to have been a death by natural causes. It had meant that the alarm could be raised and her murderer tracked down. Today, evidence-based practice requires medical professionals to read widely, complementing their practical knowledge. But in the mid-nineteenth century, practical medical and scientific experience was what defined an expert.

What the tapeworm saw

In July 1845, Taylor received more viscera to examine. It had recently belonged to 22-year-old servant Ellen Fry, who had died suddenly of sickness and diarrhoea. Her death was initially ascribed to English cholera, but when rumour got about that she was trying to procure an abortion, an inquest was opened. The post-mortem showed signs that she had been poisoned; as in the case of Eleazar Jennings, so much arsenic had been administered that it was visible inside her stomach.

Taylor found that, despite the violence of her symptoms – Ellen had died in just seven hours – there were still worms in her intestines. Some of these hardy creatures were still alive, despite being surrounded by arsenic. He found that they contained the poison, and he realised that 'when arsenic is not detected in the contents of the stomach and intestines, it may perhaps be found in the bodies of these entozoa'. He had stumbled upon forensic entomology.

Ellen had been five months pregnant, and the arsenic had not induced a miscarriage. It was a tragic case, but Taylor looked at it from a scientific perspective. Wasn't it a shame he had not been sent the foetus to examine, as it 'might have thrown some light on the absorption and diffusion of poisons'.

An entire stranger to this neighbourhood

A few months later, in October, a stranger arrived at an inn in West Bromwich in the Midlands. He stirred something from a phial into his drink, and symptoms came on within minutes. Mr Savage, a medical man, was called, and he immediately noticed a strong smell of bitter almonds after the man vomited. The stranger died within seven minutes. Savage found the phial and saw that it contained oil of bitter almonds, which is poisonous in large quantities, and he asked Taylor to make a chemical examination.

Taylor agreed that it was indeed bitter almonds that had killed the man; he performed several tests, and his colleague Aikin was roped in to identify the smell. Taylor observed that the smell persisted even when the tests failed, but that it didn't necessarily mean that the poison was still there. Yet despite their certainty of what had killed him, none of the tests could give the mystery man his name.

For he's a jolly good fellow

Christison's *A Treatise on Poisons* went into another edition in 1845, and Taylor appeared in it several times. His *Manual of Medical Jurisprudence* must have met with Christison's approval, as well as his journal articles and his editorship of the *London Medical Gazette*. To be referred to in a work by Christison meant that Taylor was now an established expert in his field.

The two men's careers mirrored each other's in some respects. They had both become professors of medical jurisprudence at young ages, but Christison had given up his chair in 1832, to become Professor of Materia Medica (pharmacology) and Therapeutics at the University of Edinburgh. Despite this apparent change in direction, Christison was still a Crown medical witness, and published more papers on medical jurisprudence than pharmacology. Christison and Taylor both led the comfortable middle-class life of professional men in capital cities; Taylor lived in a Nash terrace on Regent's Park, and Christison lived in Edinburgh's Georgian New Town.

Taylor's work had not gone unnoticed. In November 1845, he was elected a Fellow of The Royal Society. It was a mark of how respected he had become, chosen to join the eminent 'invisible college' of scientists, which still exists to this day.

Chapter 7

The Means of our Preservation
1846–47

The dreams of false philosophy

Early in 1846, Alfred Swaine Taylor, FRS, gave a lecture at The Royal Society. Not on photography, or analytical chemistry; he did not bring entrails in jars, or drawings of stomachs. It was titled: 'On the temperature of the Earth and sea, in reference to the theory of central heat'.

Anyone familiar with Taylor's paper thermometer would not have been surprised by his lecture. There were three theories, Taylor said: that the Earth is 'a sphere, filled with melted matter'; another that 'this sphere is hollow ... a mere shell'; and a third that 'the nucleus of the globe is solid, and that it is gradually cooling down to the temperature of space.' Taylor discussed how much the Earth retained heat received from the sun, and talked about experiments, burying thermometers deep within the Earth, to find the depth at which the temperature didn't alter: this, it was thought, was where the heat of the sun had no influence over temperature. The depth varied, the 'invariable stratum' being a curve, rather than a straight line. Going deeper than this, it becomes hotter, 'approaching some great source of heat, which must be situated within the interior of our globe.'

It had been observed as early as 1822, Taylor said, during analysis of Cornish mines. As the Industrial Revolution powered forwards in Britain, the issue of mine conditions became pressing: the heat encountered under the Earth's crust was not just a curiosity for gentlemen, but a practical concern for those who sent human beings into the Earth. Taylor observed that at the bottom of the shaft of one of England's deepest mines, Monkwearmouth in Durham, the temperature measured 72.6F, but 47.6F on the surface. The temperature of water rising from deep wells also showed that the Earth grew warmer the deeper you went.

And then there were volcanoes, 'fiery vents in the superficial crust'. Worldwide, they demonstrated similar phenomena and ejected similar substances, so 'it is evident that they must proceed from one common source'; that inside the planet there was 'a full red heat, extending from the equator to the poles'.

The temperature of the sea was difficult to measure; there were significant problems owing to 'the imperfection of our instruments, and their liability to be affected by unseen counteracting causes'. In 1846, no one knew how deep the sea actually was. 'Captain Ross is stated to have fathomed the South Atlantic Ocean to nearly 25,400 feet, without finding any bottom.' But Taylor did not believe that the temperature of the middle of the Earth was affecting the temperature of the sea.

In his conclusion, Taylor, like many a nineteenth-century scientist, carefully addressed the meeting of Christian faith and science. Trying to work out where the planet's interior heat had originally come from was a fruitless exercise. 'We may talk of nebulae and fire-mists, and imagine that we have determined the steps by which the great Creator proceeded to build up the planetary system, and to separate the land from the water in the globe which we inhabit; but these are the dreams of false philosophy – which admit neither of proof nor disproof.' Taylor wanted facts, things he could test, information and data. Creation, he said, could not be held up to the rigours of scientific enquiry, partly because 'our knowledge of the Earth' was sparse at the time.

He confidently claimed that the invariable stratum and the heat at the centre of the planet were good things. Ignoring the role played by the atmosphere, Taylor said that planetary heat helped to maintain a temperature that supported life on Earth. 'Hence that which some philosophers have been inclined to regard as a great element of destruction, may, in reality, be the means of our preservation.'

Obvious to the most ignorant person

The Barkers lived in Wandsworth, Surrey. Mr Barker was a coachmaster, father of several children, including 4-month-old Mary Ann. At 7.00 pm on 1 July 1846, local surgeon Mr Howell was called urgently to the Barker house; the baby had suddenly fallen ill. He arrived within minutes, and from Mary Ann's symptoms he saw at once that the baby had ingested oil of vitriol (sulphuric acid). He prescribed a mixture of magnesia to neutralise it, and two hours later, his partner George Tatham arrived to examine the child. Howell and Tatham had both seen a case of vitriol poisoning before, and recognised the symptoms. They advised on a course of treatment for the child, who responded to their care, even if her mouth was too sore for her to suckle.

Tatham had studied at Guy's; like other former pupils he turned to his old tutor for help.

In his laboratory, Taylor unpacked a box sent to him by Tatham. There was a phial, which apparently contained oil of vitriol; another, containing liquid vomited by the child; a bottle of aniseed for infants; and a damp, stained, torn napkin. Taylor's tests found concentrated oil of vitriol, the strongest kind available. The aniseed was 'a perfectly innocent mixture', which he tasted himself. 'It had a very pleasant and sweet taste,' which, fortunately for Taylor, was 'without corrosive or irritant properties.' This ruled out the acid having been placed, by accident or intention, in the aniseed bottle.

The vomit had no particular smell, he reported, but it had an intensely acid reaction. It was 'tarry or treacly-looking', the black colour being caused by decomposed blood and mucus. Taylor had seen this in other cases of sulphuric acid poisoning; before he had even read Tatham's accompanying notes, he had thought that the vomit was from such a case, recognising it on sight. He tested the vomit and the sulphuric acid he found in it was the same strength as that in the phial. Then he performed tests to see if the child had taken any medicines; he found magnesia, which Howell had prescribed. There was no other poison besides the acid.

The napkin was still damp although it had been in a dry atmosphere for some hours. This, Taylor said, was typical when sulphuric acid was present. The fabric was corroded around the stains, and Taylor snipped some fabric out to test it. He found sulphuric acid where it was damp and stained, but it was free from acid where it was clean and dry. Taylor's tests, along with the surgeons' observations of the symptoms, showed that Mary Ann had swallowed sulphuric acid. But who had given it to her?

In the days immediately following the poisoning, the little girl seemed to be getting better. Her restlessness gave way to sleep, and she was ravenously hungry. But as July wore away, it was evident that her internal injuries were too severe for her to survive. She couldn't suckle and so was 'injected' with milk and macaroni by Tatham, apparently as a nutrient enema. She was slowly wasting away, and on 26 July, the emaciated baby died. Tatham performed a post-mortem the following day, which showed the otherwise healthy baby had lost the ability to digest nourishment from the damage caused by the acid.

Three days later, a coroner's inquest was held, and Mary North, the Barker's 19-year-old servant, was found guilty of the child's wilful murder. She had been seen going into the pantry with the baby, where the oil of vitriol was kept, and immediately afterwards, the child had fallen ill. *The Monmouthshire Merlin* titled their short report of the inquest a 'warning to nurse-maids and servants', suggesting North had caused the child's death by accident.

On 31 July, Taylor went to Guildford, where North's trial would take place. The legal process was moving at alarming speed, and Taylor was lucky that a Guildford chemist gave him the use of his apparatus so that he could carry out a last-minute test. There were spots on Mrs Barker's dress, possibly caused by acid, and Taylor had to analyse them. Might it show that Mrs Barker had administered the acid to her own baby?

Mrs Barker said that the spots had appeared after she held the suffering child in her arms and the acid had dribbled from its mouth. Taylor experimented on the dress, first of all establishing what sort of dye had coloured it. Then he used pure sulphuric acid, and acid mixed with egg albumen to mimic vomit, followed by a dilution of acid. After twenty-four hours, there was no way of telling the spots apart. This meant that the spots on the dress couldn't show if the acid had come straight from the bottle or from the child. It neither incriminated nor exonerated Mrs Barker. The analysis of the dress showed 'what a wide and unexpected range chemical evidence may take on a trial for murder'. And, though he doesn't say so in the article he later wrote, how little time he had to do it in.

Mary Ann had died on 26 July. On 1 August, Mary North stood in the dock at Guildford Assizes, in the shadow of the hangman's noose. The medical evidence could not be disputed: Mary Ann had died of the effects of sulphuric acid poisoning. North's defence counsel, Mr Locke, could only alleviate her guilt by blaming someone else; the child's mother. Wasn't it true that the aniseed and the oil of vitriol were in very similar bottles? And wasn't it true that the liquid looked quite similar? Perhaps it was just a horrible accident, caused by Mrs Barker?

Mrs Barker's sworn testimony had never wavered. She had taken a cup and put some aniseed liquid in it, added a sugar cube, and then some warm water from the kettle. She tasted it herself, found nothing amiss, and gave it to her child. Elizabeth, the Barker's 11-year-old daughter, drank the rest of the cup's contents, and Mrs Barker passed the baby to North. Mrs Barker went into the parlour, and a few minutes later, ran back into the kitchen, alerted by Elizabeth, who exclaimed, 'Mary, don't hurt baby.'

North approached Mrs Barker, the child in her arms, saying, 'Oh, dear me, ma'am, what is the matter with baby?'

Locke asked Tatham to perform an experiment in court, to show what would have happened if Mrs Barker had used acid, instead of aniseed, in her mixture. Rather than use aniseed, Tatham was directed to put sulphuric acid, then sugar, then water, in a cup. The acid carbonised the sugar, creating 'a black fluid, at a very high temperature'. Surely this experiment showed that Mrs Barker could not have accidentally given her child acid instead of aniseed? It was 'obvious to the most ignorant person', Taylor thought.

Locke made Tatham perform another experiment, adding the ingredients in a different order, and to use double the amount of water. So Tatham added the acid, then the water, and finally the sugar. There was little sugar left, and by adding the water to the acid first, it saturated the acid, so it couldn't carbonise the sugar. The result was 'a deep yellow fluid, of intense heat, and which gradually became darker on cooling'. Locke held this up as evidence that Mrs Barker had accidentally poisoned her child.

Taylor was dismayed; what Locke had actually proved – twice – was that Mrs Barker could not possibly have made the error. If the mixture given to the child could not have carbonised the sugar, then it could not have carbonised the child's insides. The acid must have been administered another way, and North's skulking in the pantry was surely the clue.

There was even more reason for Taylor's consternation when Locke gave a closing speech that made claims about medical and chemical evidence that he had not cross-examined the expert witnesses on. He made much of the fact that Taylor had been unable to smell aniseed in the child's vomit, but Locke had not asked Taylor about this. Had he been asked, Taylor would have explained that the time taken for him to receive the evidence, and the smell of the carbonised 'organic matter' in the vomit would have overwhelmed it. Locke instead said that the lack of aniseed smell meant the child had not been given aniseed by the mother at all, but oil of vitriol, by accident. With little deliberation, the jury returned a verdict of not guilty.

Matters were not improved when John Harrington, an attorney on the defence team, bragged about the cleverness of the courtroom experiments in the pages of *The Lancet*. The editor praised him for saving 'the life of the innocent accused person' by chemistry, not 'by the eloquence of the counsel'. Nothing could have riled Taylor more, unless it had come from the pen of Henry Letheby.

Taylor wrote to Harrington, to 'call your attention to a most serious mis-statement' in his letter to *The Lancet*, explaining why he thought the experiment was ridiculous. 'In the course of now sixteen years' experience, I have never met with a grosser perversion of chemical evidence in a court of law than that made by Mr Locke, apparently at your suggestion.' Harrington replied, marking the letter 'strictly confidential', so Taylor could not share its contents with the world – but he sent a copy of his reply to Harrison to *The Lancet*, which duly printed it, taking the opportunity to explain how he stood on the case.

Taylor was angry about the 'perversion of chemical evidence' but was also defending the reputation of his former pupil, Tatham; 'the unprejudiced evidence of a respectable surgeon has been severely and unjustly censured.' Taylor explained that the results of the experiment performed in court 'conclusively proved that the prisoner was guilty!'

Sometime in October, *The Lancet* received a reply from Harrington, but they did not print it. By then, Taylor had written a forty-six-page article in *Guy's Hospital Reports*, which put the matter to rest. He felt that the evidence showed that Mrs Barker had not accidentally poisoned her child; it must have been Mary North, but he said that it would have been up to the jury to decide if it was murder or an accident. After all, there had been no motive for her to kill Mary Ann. And so, the Wandsworth Vitriol Mystery remained.

The Essex Lucretia Borgia?

A couple of weeks after North's acquittal, Taylor received a collection of items from the village of Clavering in Essex: a home medicine chest and the viscera of two young boys, who had been dead for nineteen months.

Gossip and rumour swirled about the village. A woman called Sarah Chesham had apparently been poisoning the illegitimate son of maid servant Lydia Taylor. Lydia claimed Sarah had been paid by the child's father, a local farmer called Thomas Newport, so that he would no longer have to financially support the baby. Sarah was remanded in custody until the next Assizes, on a charge of attempting to poison; Lydia's child yet lived. The locals remembered the sudden deaths, the year before, of two of Sarah's sons, James and Joseph; the symptoms had not been unlike arsenic poisoning. The little boys, buried together in the same coffin for economy's sake, were exhumed. Where chitchat and hearsay reigned in the village, in his laboratory Taylor would invoke reason.

Taylor found arsenic trisulphide inside the boys; what was once white arsenic had been yellowed by the bodies' decomposition. There was a lot of it, taken during life. In Joseph's viscera, Taylor found eight to ten grains of arsenic – enough to kill an adult. James's contained even more – twenty to thirty grains, which would have killed three or four adults. Considering this was what was left after some had been absorbed into the body, and after the children had been violently ill, then the original dose would have been even larger. Their vomiting had been so copious that it had dripped through the floorboards into the cottage below, leaving their neighbour crying as she mopped up the mess.

Taylor made the journey up from London to the village pub, where the inquest was held. He explained his findings and, ever one for visual demonstrations, showed slivers of the boys' stomachs on glass slides, and pieces of copper gauze from the Reinsch test. His evidence satisfied the coroner, Charles Carne Lewis, that Joseph and James had died of arsenic poisoning but it would be up to the jury to decide if their mother should be sent to trial for this crime too. The coroner requested Taylor keep these artefacts safe, and 'the learned professor', as one of the newspaper reports described him, went back to London.

At the end of September, Lydia Taylor's baby died. The charge of attempted poisoning suddenly became more serious. Attempted poisoning carried the punishment of imprisonment or transportation, whereas murder carried the death sentence. Now the child was dead, Taylor could look inside for signs of poisoning. Perhaps he could find out what was going on amongst the thatched cottages of East Anglia.

As it turned out, perhaps nothing was amiss at all with the baby's death. Taylor could find no signs at all of poisoning; 'there was nothing to form an opinion that death had been caused by poisoning either directly or indirectly.'

And what about Sarah Chesham's medicine chest? *The Times* had already found her guilty, in a hair-raising editorial. They called Sarah a poisoner-for-hire 'whose employment was as known as that of a nurse or washerwoman – who could put any expensive or disagreeable object out of the way, and who, as it was understood, had practised her infamous trade on her own children.' Before Taylor's analytical report had been made public, *The Times* claimed that Sarah's medicine chest consisted of 'an assortment of poisons – ointments, powders, and the like – such as was discovered by Claudius in the private cabinet of Caligula'.

But Taylor's report showed that her medicines were prosaic; common items in many Victorian homes. Although three were poisonous, they were 'not of a dangerous character unless administered in large quantities'. Lydia had claimed that she had seen Sarah smearing a pink salve over her baby's mouth, but nothing that Taylor found in the medicine chest matched the description. Lydia had reported only a couple of occasions when Sarah had been around her baby, which was not often enough for a case of slow poisoning, requiring many tiny doses over a long period. It cast doubt on the grounds for Sarah's original arrest.

Just before Lydia had come forward with her accusations, the peculiar arsenic poisonings of a family in Happisburgh, Norfolk, had been in the news. Jonathan Balls, an eccentric elderly man, was exhumed, and found to have died of arsenic poisoning. Nine relatives of his, who had died suddenly, were also exhumed and five of them were found to contain arsenic. Had this suggested poisoning to Lydia? She had said that an apron she lent to Sarah had 'dropped to holes', which might imply that Sarah had used an acid to poison the child – an idea perhaps suggested by the Mary North case.

In October, while Sarah Chesham sat in Chelmsford gaol, Taylor published an article in *Guy's Hospital Reports*. It was nearly 100 pages long, covering cases of poisoning by oil of vitriol, calomel, arsenic, corrosive sublimate, lead, bitter almonds, and prussic acid. There were poisons everywhere, used by murderers and suicides alike. And there were the deeply unfortunate who were accidentally poisoned by their doctor.

Taylor had been referred a case by magistrates in Reading who were investigating the death of a man who had been prescribed calomel – mercurious chloride – for a kidney complaint by a quack doctor. The man's wife was illiterate; unable to read the medicine label, she had accidentally overdosed him. Taylor tried to analyse what was left of the medicine, to work out the pill's strength, but the calomel dose varied from pill to pill, and contained pieces of a green leaf that science at the time could not identify.

A pressing medico-legal question was whether calomel should have been prescribed in the first place. Other medical professionals saw no problem, but the case, Taylor believed, showed 'the injurious effects which may occasionally arise from the use of even small doses of mercury in disease of the kidneys'. Taylor had sought the advice of his peers; he added at the end an extract from a letter he had received from Christison. The eminent Scotsman hadn't observed any problems with calomel prescriptions, but 'we watch it more narrowly' in Edinburgh; anyone suffering an accidental overdose would be spotted before fatal symptoms set in.

It wasn't only scientists who wrote about poisons. Then, as now, poisons furnished the weapons for characters in works of fiction. Edward Bulwer-Lytton's novel *Lucretia* was published in November 1846. The story of a pair of poisoners, it was criticised partly for the questionable quality of its writing, and partly for its content, which, it was feared, might inspire copycat poisoners. The main character was called Lucretia Clavering, deliberately named after Lucretia Borgia, the infamous Renaissance poisoner, and there was exciting contemporary resonance courtesy of the village of Clavering, home to Sarah Chesham.

Taylor was the expert witness in the Chesham cases, so it is hardly surprising that he should be very unimpressed by Bulwer-Lytton's literary creation. He departed from the usual content of his *London Medical Gazette* editorials, about the weather's effect on health, or the dangers of burials in London, and instead wrote about 'the crime of Secret Poisoning'.

His first editorial to address this appeared on 15 January 1847, where he wrote that it was his calling and that of every 'enlightened medical practitioner … as a man of science, and as a citizen of the world' to limit 'the progress of those various causes which are known to produce death, either by accident, design, or by gradual decay of the powers of life'. Secret poisoning was 'a great moral evil', which had, he believed, 'for several years past been upon the increase in this country, and which has, throughout its progress, been evidently marked by consequences of so fatal a character, that it loudly calls for the interference of society at large, and most particularly of the members of our own profession.'

What could explain this increase? Partly, Taylor believed, it was the way that inquests and trials were reported in the newspapers, which 'have laid open to the popular view nearly all the means of detecting homicide with which medical jurists are furnished' and 'have rendered particularly apparent the various means of escape'. Taylor claimed this had led to 'the less rational members of the community' coming out in favour of those accused of poisonings, who had 'the air of popular martyrs' rather than 'the aspect of wretches whose souls are branded with the crime of secret and unavenged murder'.

If that wasn't bad enough, poisonings appeared in popular romances. At the time, crime fiction focused on the felon, not on detectives, and criminals were often portrayed as exciting rogues. Crime is ugly, until it is 'decorated by the false lustre which the imagination of a romance writer can throw over things in themselves the most disgusting'. And so Taylor made an announcement. 'We do not hesitate to declare, that the fictitious literature of France and England, in which the most atrocious offences have of late years been printed with exaggerated details, and in the most brilliant colours which an excited imagination is so well able to give to the unnatural, has been mainly instrumental in producing the recent increase of the crime of poisoning in this country.' Long has popular culture been blamed for society's ills.

'A large proportion of modern novels may be regarded as convenient hand-books of poisoning,' Taylor went on. 'In every page are detailed the means whereby secret murder has been successfully perpetrated, with suggestions on the cautions to be pursued by future experimenters upon the lives of others, in order to avoid detection.' He didn't name *Lucretia* or her creator, but Taylor referred to 'the most eloquent novelist of the day' who 'has just given to the world a work of fiction, the entire plot, details, and moral (?) of which form a most complete revelation of the art of murder by poison.' That bracketed question mark could not be more sarcastic.

Taylor included a translation from a recent French novel, which went into technical detail about how to poison someone and hide it: 'The sole object of the writer is to encourage murder, and baffle the power of the law.' He called on his colleagues in the profession 'to mitigate the effects of this evil, and prevent its extension' but gave no idea as to how this might be achieved, short, perhaps, of banning such novels from their own homes.

A fortnight later, in another editorial, he returned to the negative effect of certain literature on the minds of the populace. Trials for poisoning had been on the increase, as were publications on it, and Taylor, presumably confusing cause and effect, believed that these writings 'have been strongly instrumental in rendering the crime in question so frightfully prevalent as it at present is'.

Felon literature, 'the great object of which appeared to be to give an air of chivalry and romance to the violent and nefarious actions of the highwayman and the pickpocket, the burglar and the prison-breaker' had been shown, Taylor stated, to have increased certain kinds of felony. Therefore, novels about fictitious poisonings, and book-length write-ups of poisoning cases for the general reader, were surely to blame for an increase in those crimes. What about poet and author Letitia Elizabeth Landon, who had written about someone being poisoned by prussic acid, and who had ended up dead from it herself?

And now, from the great quantity of literature available on poisoning, the public were able to refine their craft, and had learnt the art of slow poisoning by arsenic. 'One atrocious case of this kind, i.e. of slow poisoning by arsenic, which completely deceived the medical attendant, is now waiting for trial in this country. In order not to prejudice the accused, we cannot at present make any further allusion to it,' Taylor said.

Just after the Lafarge case 'had been circulated from one end of the country to the other', a man called Hitchins, in prison for abducting voters during an election (an extraordinary and yet not uncommon crime at the time), had been sent a plum cake laced with arsenic. That same cake was probably the one that Taylor kept in his laboratory. Hitchins had not been killed, but whoever had sent the poisoned cake to him was never found. Taylor was specifically linking them; had the press not whipped themselves into such a frenzy over the Lafarge case, perhaps whoever had tried to poison Hitchins would not have come up with their terrible scheme.

Taylor complained about the number of publications about servants avenging themselves on the families they worked for. He gave examples of two cases where cooks had killed people by putting arsenic in food. They had, he suggested, been influenced by reading about the cases of other servants that had been printed in 'cheap periodicals'.

Once a case reached trial, it was bedevilled by judicial failings; 'the prosecution is often left in the hands of parochial "Dogberries", who go through the forms of law at as cheap a rate as possible.' Barristers then pervert 'medical evidence, like that displayed on a recent trial for murder at the Guildford Assizes', Taylor wrote, the pointless witness box experiments still smarting.

Bulwer-Lytton's *Lucretia* had been attacked on many fronts: it had received a sarcastic drubbing in *The Times*, as well as in many local newspapers. So the beleaguered novelist wrote *A Word to the Public*, a pamphlet to defend his work and himself, and he sent Taylor a copy. He may not have expected to receive a review in the *Gazette*, but Taylor was unable to resist. The review was six pages long, but in some respects cannot really be called a review at all. Taylor thundered on in his furious, standard-bearing manner, shredding *Lucretia* and all works like it. He softened slightly, saying, 'True it is, that the author, in almost every page of his narrative, expresses his detestation of the malefactor, and his abhorrence of the crime.' Taylor may have realised that attacking Bulwer-Lytton personally was dangerous territory. 'Yet why,' he went on, 'should the secrets of a long-forgotten art of assassination be revived, and laid open to the public eye? Can it be right that guilty and designing imaginations should be fed with such materials?'

Taylor praised the author for his previous work, but this 'incurs a fearful responsibility'; by writing about immoral deeds in *Lucretia*, he had misled readers, and had 'almost conferred an air of sanctity' on his errors. Books like *Lucretia*, Taylor said, 'impart to the basest crimes a dignity which never actually pertains to these atrocities.'

If it all seems over the top, as if Taylor was paranoid and thought poisoners lurked in every shadow, then bear in mind that he was the man who was often sought when someone died of a suspected poisoning. He had to test human viscera for poison, and when he found it, it could well mean that someone had been murdered. He knew that poisons were too easily accessible, and as the fight for regulation was a difficult one, he could at least challenge the way that poisoning was presented in print. He had seen Thomas Jennings in the dock, the man who was found guilty of poisoning one of his own children, and had perhaps killed another. Taylor spoke from experience when he said, 'Nothing heroic belongs to the cowardly, remorseless, crafty, shrinking spirit of the murderer by poison.'

My mother has been a good mother to we

Sarah Chesham had spent six months in a cell by the time of her trial in March 1847. At the time, defendants stood trial for one offence at a time, so she was tried first for the murder of her son, Joseph. If she was found guilty, she would not stand trial for the others; there would be no point, as she would be hanged. Taylor reiterated his evidence in court: the boy had most certainly died of arsenic poisoning. Sarah's defence counsel, Charles Chadwick Jones, could not find a way around that, so his 'forcible and eloquent' speech focused on what a kind and loving mother she was; she could not have killed her own son. The jury didn't take long to reach their verdict. They deliberated for ten minutes and found her not guilty. 'We have no doubt of the child having been poisoned, but we do not see any proof of who administered it.' As ever, the query over administration prevented a conviction.

Cleared of one murder, she was now tried for the murder of her other son, James. Philip, one of her surviving children, backed up the image created by her defence counsel by saying, 'My mother has been a good mother to we.' There was the medical evidence to go through again. The chain of hands through which the viscera had passed to arrive in Taylor's laboratory had to be established, to avoid the imputation that the evidence had been tampered with. Unfortunately, Professor Graham, Taylor's colleague who was part of the chain, was on his way home to London. Someone was sent from the court to Chelmsford station to send a telegraph message down the railway line to urge his return. Luckily, the message reached Shoreditch before Graham's train did, and the embarrassed professor made his way back on the northbound train. With its trappings of modern technology – toxicological analysis, trains and telegraph messages – the trial was turning out to have elements of the Tawell case about it, but Sarah Chesham was once again acquitted.

The following day, Sarah Chesham was tried on several indictments related to the illness and death of Lydia Taylor's baby. His mother cried bitterly as she gave her evidence, insisting that 'Sarah Chesham has been the death of my child and nobody else.' Had there been little confidence in medical evidence, Sarah might have been in trouble, but Professor Taylor cleared her. No matter what Lydia claimed, there was no poison found in the child's remains, and no damage that could be ascribed to poison. The judge threw out the case before the jury were able to deliberate. She was a free woman … for now.

Chapter 8

The Only Friend I had in the World
1848

All he knew about the matter

In May 1848, Taylor's expertise was called on to investigate the death of another illegitimate child. Beatrice was the daughter of Hannah Bowyer, who lived in Haverhill, Suffolk. It was not prussic acid or arsenic that Hannah was accused of having used to kill her own child; it was hemlock.

Hemlock grows in most places in Britain, and its leaves resemble cow parsley. Hannah said she had picked some hemlock to make a tea to pour on Beatrice's swollen ankles. The child was two years old, and had always been sickly. Hannah had a new man in her life, William Glasscock; it might have been assumed that Beatrice was an encumbrance as they shuttled back and forth between the workhouse and their cottage. Beatrice developed a bowel complaint, and Hannah took her to see the local surgeon. His assistant prescribed some medicine, and Beatrice died very soon after taking the first dose. The villagers knew that Hannah had picked hemlock; wanting to clear away the suspicion surrounding her, Hannah 'particularly wished for the examination and inquest'. The surgeon's post-mortem didn't show the work of any poisons, and the coroner directed the jury to give an open verdict.

But the following day, William came forward. He 'stated that he could not rest until he had told the Jury all he knew about the matter,' and gave a long statement, which 'completely established a case against Hannah Bowyer, principally on her own declarations'. Thanks to her fiancé, Hannah was arrested, and would have to wait until August to go to trial. Taylor was sent Beatrice's viscera and a description of her symptoms. It was said in Haverhill that Hannah had declared she would poison her child. Would she be next to dangle from Calcraft's noose?

I never hurt a hair on his head

Then, in June, suspicions of another arsenic poisoning in Essex came to light. William Constable, also known as Spratty Watts, had lived in Wix, a village in the north-east of the county. He was a widower who boarded with his half-sister, Mary May, and her family. William drove sheep and dealt in scrap metal, working in the fields when he could. He had been arrested earlier in the year on suspicion of stealing from Reverend Wilkins, the vicar of Wix, but was released.

Mary had found out from her neighbour, Susannah Foster, about the burial club in Harwich. For a regular payment of mere pennies, a loved one could be ensured a good funeral, with a payout of £9 or £10 – about half an agricultural labourer's annual wage at the time. It meant a respectable send-off, and the avoidance of the medical school anatomy class. Even though there was a club much closer in Great Oakley, Susannah had signed up her own brother in the Harwich club, and Mary signed up her brother too. Without his knowledge, William would be sure of a good funeral, which was as well: less than a month after Mary signed him up, he died.

The club initially agreed to pay for the funeral, but Constable had a pauper's burial, paid for by the parish. Either Mary had suppressed their agreement and hoped to claim the money for herself, or the club immediately reneged on it. After the swift funeral, Mary pestered Reverend Wilkins for a signed letter saying that her brother had been in his usual good health when she entered him in the club. Wilkins didn't like Mary; she had been a widow when she arrived in Wix, and she had, so he claimed, lived unmarried with the man who would become her second husband. They said they could not afford the cost of marriage, so Wilkins married them for free. But does living in sin lead to murder?

Wilkins wasn't taking any chances, so he called on Inspector Raison of the Essex Constabulary to investigate. When Raison asked Mary if she had bought any medicines for her brother when she went to Harwich, she bridled at him. 'Oh Lor', Mr Raison, you think I got something and gave him, and killed him, I know that's what you think. Poor old soul, I never hurt a hair on his head, he was the only friend I had in the world.'

Constable's symptoms – vomiting, diarrhoea, stomach pains, death three days after the illness came on – could easily have been English cholera in the warm summer, but, with the lure of burial club money, his death took on a more sinister aspect. There was precedent: across the country, from Yorkshire to Somerset, people had been hanged for poisoning family members for burial club money. An inquest was opened on 30 June, and, as the local surgeon wasn't able to perform the toxicological analysis himself, Constable's viscera were sent to Taylor.

Taylor saw a yellow streak in Constable's stomach, which looked a lot like arsenic trisulphide. He performed the Reinsch test and proved that it was, indeed, arsenic. When he tested another part of the stomach and the duodenum, he found arsenic present there, too. Mary was arrested. On 7 July, Taylor attended the inquest meeting in Wix. The schoolroom had been requisitioned because the pub, where the inquest had been opened, had soon overflowed with curious locals.

Taylor confidently gave his evidence. 'I have not the slightest doubt whatever, that death was caused by inflammation of the stomach, and that the inflammation was caused by the action of arsenic.'

By the end of July, Mary was on trial at Chelmsford Shire Hall, where Sarah Chesham had stood a year before. Her defence counsel, Charles Chadwick Jones, was the same man who had successfully defended Chesham. Would he save the neck of another Essex woman accused of arsenic poisoning?

Mary had got herself into a muddle. The evidence regarding the burial club was damning, and she even admitted to having bought Butler's Rat Poison, which had an arsenic base. But it was possible that her brother could have been poisoned accidentally. On the day he fell ill, he had been droving sheep, and it was well known that arsenic was used for treating wool. Mary said she had given her brother some ginger beer as he left and, as Taylor had written back in 1841, the bottles used for ginger beer were sometimes recycled, having once contained arsenic.

But Mary had asked people to lie for her, and had lied herself, claiming that her brother was suicidal. 'As true as God, child,' she claimed, 'he would have hung himself one day not long ago if it hadn't been for me; he took a line, went up the field, and I ran after him.' No one else could corroborate her statement.

Try as he might, Jones could not extricate his client, and Mary May was found guilty, with a recommendation to mercy. Whether the jury entertained some slight doubt, or they objected to the death sentence, is unknown.

The judge was Frederick Pollock, a friend of Taylor's. Pollock's sons were keen photographers, which may have had something to do with their 'terms of intimacy', as it was later put. Not feeling merciful, Pollock put on the black cap and sentenced Mary May to death. 'You appear to have been actuated merely by this sordid love of a small, an exceedingly small, sum, and for this you have destroyed the life of a near relative, and periled your own soul.'

Roughly made as by a cottager

A few days later, Taylor was in Ipswich; Hannah Bowyer, the Haverhill hemlock poisoner, was on trial.

Hannah's local paper was not very complimentary about her. She 'was of small stature, and of rather a repulsive appearance.' When she pleaded, Hannah said 'in a firm voice, "I am not guilty, I done it innocently."' Her Suffolk twang was reported true to life.

Hannah had dictated a confession to the landlord of the inn where she had been taken after her arrest. She confessed that 'Glasscock persuaded me to poison my child.' She described picking some 'sheep's parsley' but Glasscock had said, 'that would do as well as hemlock.' She had boiled it in a brass saucepan, following Glasscock's guidance; he told her that half a cupful would kill Beatrice. She gave it to the child, and 'when we went to bed he said, he was heartily sorry for what had been done, and I said if there was any punishment I would bear it myself. Glasscock said he was d—d glad.'

There was evidence from locals who alleged Hannah had told them, or implied, she would do away with her child. One woman, Elizabeth Nunn, treated the court to an expletive-laden quote, which she alleged Hannah had said over the ailing Beatrice: 'I'll be d—d if that it will live till night, and be d—d if I want it should; d—d if I don't hate the child.'

Defendants were not supposed to give testimony in court, and maintained silence during the trial, other than to whisper to their counsel. But Hannah objected to Elizabeth's words, and shouted from the dock, 'Betsey, you are a false woman; I stand here without a friend in the world; you may well look pale, you may; you will go to the devil, die when you will; you will go to a rare place when you die.'

Yet Hannah was not as friendless as she thought. The tall, imposing scientist from London – 'the well-known chemist', as one newspaper described him – and his professional peers, had yet to give evidence.

The three local surgeons had found no evidence that Beatrice was poisoned. One had diagnosed scrofula in the child, and thought that she might have died of a disease of the mesenteric glands. Another had made a hemlock tea, following the instructions William Glasscock had given Hannah, and fed it to a rabbit. Although 'I expected to have found it dead' he instead found 'that it jumped about as lively as ever'. The tea smelt strong, but the odour was absent from the child's remains.

And now it was time for Taylor. He had found 'a slight patch of redness on the stomach, but no appearance of ulceration to account for death'. There was some undigested food in the intestines, which was free of vegetable matter – 'I found no trace of leaf, or root, or stalk, or anything of the kind' – and it was free 'from any particular odour'. Neither was there any odour in the stomach. It was possible that the poison had been absorbed, but if death was sudden, then absorption could not have taken place. The symptoms described could have indicated poison, but could equally have been the result of natural causes.

The judge, Baron Parke, appears to have been irritated by the evidence being used against Hannah. He interjected several times during Glasscock's testimony, saying, 'Glasscock, mind what you are about!' and 'You know it's your duty to speak the truth, mind that.' He seems to have been under the impression that there was a conspiracy against the defendant. He intervened and asked Taylor, 'Supposing these gentlemen are correct in saying that two days after the death a careful examination was made, and no trace of poison found, do you conclude from this that poison had not been administered?'

'Certainly,' Taylor replied. 'When roughly made as by a cottager I should expect to find traces of it, if poured out of a cup, at the bottom of which were leaves and vegetable matter, and should expect to find similar appearances in the stomach.' As he hadn't, then Taylor was of the opinion that Beatrice had not had hemlock administered to her at all.

Hannah Bowyer was found not guilty. Her salty tongue continued to wag; when she saw the turnkeys come back into court, she was heard by one reporter to say, 'There come the devils.'

Her trial showed that even if every piece of circumstantial evidence pointed to guilt, even down to a confession, scientific evidence could still save a defendant.

Taylor's evidence might have rescued Hannah Bowyer from the scaffold, but it had doomed Mary May. The Quakers had got up a petition to save her from execution, but Pollock was unmoved. Sir George Grey, Secretary of State, had to decide on such petitions, but he refused to pass Mary's to the monarch, the only person who could grant clemency. Grey said 'it was one of the worst cases that had come before him', and in mid-August, Mary was hanged outside the gates of Springfield gaol by William Calcraft. He was from nearby Little Baddow, so it was a homecoming for him.

It was to be a treat for her

At the end of August, Taylor was at the Old Bailey. This was not a poisoning from a labourer's cottage, but involved the servants of aristocracy. Mrs Dore's father had been the butler and steward of Princess Sophia, Queen Victoria's aunt, and her husband had been the outrider of William VI, continuing in Queen Victoria's service; by 1848, he was the butler of a county magistrate. Mrs Dore and her mother, Mrs Spry, lived in a house together on Lower Grosvenor Place in London. The Dores' first child had been born a month after their marriage in 1846, but the little girl lived only seven weeks. Another daughter, Mary Ann Theresa, was born in March 1848. Apart from a bout of thrush, she seemed quite well.

In July, Mrs Dore was heading to Ryde on the Isle of Wight on a trip with her friend Harriet. Mrs Dore's husband had been born on the island, so she may have planned to visit some of his family. 'It was to be a treat for her,' Harriet said, but it would turn out to be anything but.

At two o'clock in the morning on the day they were to leave, the baby woke up crying in pain. It seemed to have been caused by the tooth she was cutting, so at half past eight, Woolmer, the surgeon who usually attended the family, was sent for. He was out of London at the time, so his assistant, Hustler, visited. He didn't think her condition was serious, and dispensed calcined magnesia, aniseed water, syrup of poppies, and a powder of calomel and sugar. He saw no reason why Mrs Dore shouldn't go on holiday.

While on the steamship across the Solent, Mrs Dore went to change her daughter's nappy. She found that she had passed bloodied stools, and the crying of the child continued. They arrived in Ryde and went to their lodging house, where another surgeon was called. Bloxham thought the child had intussusception, where the bowel causes a blockage by telescoping back onto itself. He 'tried to explain it, by passing one hand into another.' He didn't hold out much hope for the child, but prescribed more medicine anyway – mercury and chalk, castor oil, and a mixture of carbonate of soda and citric acid. Mrs Dore was in great distress and wanted to go home. Having lost one child, she feared she was about to lose another.

Their Isle of Wight holiday cut short, the women arrived back in London. The baby's condition had noticeably deteriorated, and Woolmer was called again. He still had not returned, so Driver, a surgeon who took on Woolmer's more difficult cases in his absence, paid a visit. He prescribed more castor oil, and a mustard poultice to be applied to the stomach. He came back two hours later. 'There was no relief from anything I gave it – I could see it would not recover, and that most likely it would not live till morning.' He directed Hustler to give the child calomel and, the pharmacopoeia being all but exhausted, applied a leech to her temple.

Four-month-old Mary Ann Theresa Dore died at about three o'clock that morning. Driver suggested a post-mortem, but mother and grandmother objected. Mrs Dore did eventually agree; she had initially refused due to Driver's rude manner, but when Mrs Spry saw Driver, 'she was very angry, and desired me to leave the house.' The post-mortem was carried out by Woolmer, and Bloxham, the Isle of Wight surgeon, was proved right; the child had indeed died of intussusception. It accounted for all of the child's symptoms, 'the sickness, the cries, the bloody evacuations, the convulsions'.

It would have ended there, but Woolmer, who had headed for the intestines first, decided to check the child's stomach as well. He remembered her bout of thrush, and he wanted to see if it had caused any damage. He found a black patch in the stomach, and so he sent it for examination to Julian Rodgers, analytical chemist and lecturer. And Rodgers found arsenic.

An inquest was opened at the end of July, and Mrs Spry and Mrs Dore, 'two very respectable looking women, attired in deep mourning', were both present. At the end of that first meeting, they 'left the dock declaring their innocence'. But the next and concluding meeting would not end well for them. Although Bloxham had travelled up from the Isle of Wight to give evidence, saying he didn't think the child had been poisoned, Rodgers's chemical evidence exerted a mesmerising pull on the jury. He insisted that, from the appearance of the stomach, the child must have 'taken the arsenic at least two days before death'. But how could this be, when Mrs Spry denied having it in the house, and Mrs Dore said 'she had never seen arsenic before in her life'? The jury took two hours to deliberate, and they came back with the verdict: that the baby had been poisoned with arsenic, 'with the guilty knowledge of the persons in whose custody she was'.

Dignified in adversity, Mrs Spry exclaimed, 'I hope the Lord will forgive you all, gentlemen.' Mrs Dore dissolved into tears. They were both sent to Newgate prison, but were bailed two days later.

Popular opinion was in their favour. A newspaper in Essex, a county well used to arsenic poisonings by now, reported that the coroner's jury made their verdict on 'no very definite evidence'. But in the middle of August an inquest opened into the deaths of the Dores' first child and Mrs Spry's 11-month-old son. Both infants were exhumed from St George's in Bayswater, and were identified despite the crowded conditions of the churchyard. Rodgers carried out more chemical tests, but found no evidence of poison, and a surgeon who had attended on the babies said one had suffered from diarrhoea, and the other from convulsions. With no chemical evidence, the inquest jury returned a verdict of natural death. But the women would still have to stand trial for the murder of Mary Ann Theresa.

On 22 August, Mr Clarkson at the Old Bailey 'said he had to make an application of rather an unusual character.' Mrs Spry and Mrs Dore were asking 'that the court should make an order that Dr Taylor, the eminent chemist' should make an analysis. Taylor's 'skill in these matters was well known' – the women selected him for a second opinion, one that might save them from the noose.

Taylor, who had no doubt been watching the case unfold in the newspapers, had already been contacted. Clarkson had a letter from him, saying that he could not make a satisfactory examination unless he was able to use his own laboratory at Guy's. The recorder of the court said it was an unusual request, and he wasn't sure that the court had jurisdiction in the matter; but 'in furtherance of public justice', he agreed. He stipulated one condition, however; that the professional men who already had possession of the stomach should be present when Taylor made his analysis.

The trust that legal professionals had in Taylor is clear when Clarkson said 'that the character and position of that gentleman were sufficient guarantees that nothing improper would be done'.

The recorder agreed; he 'was quite aware that Dr Taylor's evidence would be above suspicion.'

On the day of the trial, Taylor listened to the testimony given by all the witnesses – the several medical men who had examined the child and prescribed for it, the child's aunt who had seen its illness, and its mother's friend who had travelled with it. By the time Taylor stood to speak, he had combined the verbal testimony and his own analyses into a devastating narrative that knocked Rodgers's evidence for six.

Just like Rodgers, he too had found arsenic in the child's stomach, but that wasn't what had killed her. He disagreed with Rodgers, who said the arsenic must have been there at least two days before her death; Taylor believed it was more likely to have been eight or ten hours. Rodgers claimed that the intussusception had been caused by arsenic, but Taylor said this was 'quite unfounded'. None of the witnesses had described arsenic symptoms, except for vomiting, but that was common to both intussusception and arsenic poisoning. Taylor was of the opinion that the arsenic had not entered the child's stomach until it had returned from the Isle of Wight. He explained that arsenic has an 'exceedingly powerful' effect on children; the effects would have been immediate. The child's pain came and went, which was a symptom of intussusception, not arsenic poisoning. 'No person of ordinary science could possibly mistake the effects of arsenic,' and none of the many medical men had identified arsenic symptoms in the child.

Taylor had tested the stomach and its contents for calomel, but found none present. Although he didn't spell it out, he implied, as did the questions of the barristers, that one of the medical men had slipped up; one of them had accidentally given the child arsenic instead of calomel, which, like arsenic, was a white powder. But it wasn't for Taylor to investigate at the trial. It was enough that his confident, self-assured evidence showed that the child had died of natural causes, and had not been murdered. Mrs Spry and Mrs Dore were acquitted.

The case demonstrated two things that caused unease: firstly, that the child could have died by calomel being muddled up with arsenic; secondly, it showed that experts did not agree. Having been asked to appear on behalf of the prisoners, perhaps Taylor considered the evidence in their favour. But it could simply be that Taylor had more experience than Rodgers. Taylor began his evidence by declaring that he was 'eighteen years lecturer on Chemistry and Medical Jurisprudence at Guy's Hospital'; he wanted the court to understand that he spoke from a position of experience and authority. Even so, he and Rodgers had similar careers – both were surgeons who had an interest in chemistry, yet their interpretations differed. Rodgers's idea that arsenic poisoning caused intussusception was a stab in the dark, and Taylor, with his personal experience, and his encyclopaedic knowledge of arsenic poisoning cases, did not agree with him. In the end, Taylor had the more convincing argument.

The horrors of the modern school of French novelists

The rumours in north-east Essex would not be quieted. Apart from newspapers claiming, somewhat implausibly, that Mary May had murdered sixteen of her own children, they were also intimating that arsenic poisoning on a grand scale was going on in the area.

The next woman to be arrested was Hannah Southgate, whom the press incorrectly claimed to be Mary May's sister. Hannah also lived in Wix, but her crime was alleged to have taken place the year before in the village of Tendring, a couple of miles south. It was said by the gossips that she and her first husband, Thomas Ham, had lived together unhappily; that she had drunk a lot, and had been carrying on with a farmer called John Southgate. When Ham died, his symptoms being rather similar to those suffered by William Constable, Hannah had married John with indecent speed. An inquest was opened.

On 25 August, Taylor performed his analysis. Ham had died in April 1847, and the stomach contents were 'in a partial state of decomposition'. Under a microscope, he could see that the partially digested meat in Ham's stomach was 'completely penetrated by the particles of arsenic', mixed with something like sand, which wasn't unusual in commercial arsenic-based rat poisons. He found arsenic in the liquid contents of the stomach, the coats of the stomach, and in the middle of the small intestine. Red patches in the stomach and intestines showed irritation in life; Ham had taken the arsenic before his death.

Taylor sent his evidence to the coroner, and Hannah was arrested. At the inquest, Hannah's solicitor desperately suggested that the arsenic Taylor had found had been absorbed during Ham's sixteen-month interment from a body in an adjoining grave. Taylor replied, 'I never met with such a case.' There was laughter from the audience at the expense of the legal expert, made a fool by the man of science.

Hannah, presented as a licentious drunk, and possibly a murderer too, was sent to Chelmsford to wait several months for her trial. *The Times* compared her to Madame Lafarge, and wrote that a wicked system of poisoning existed in Essex, 'more terrible than the horrors of the modern school of French novelists', neatly echoing Taylor's editorials on 'the crime of Secret Poisoning'.

'A couple of adventurers', according to the *Essex Standard*, who had hoped to be paid 'two-pence per line for their compilations', wrote a sensational and somewhat inaccurate piece that tried to lift the lid on the poisoning 'system' in north-east Essex. They printed gossip culled from locals, which allowed them to print the names of people suspected of having been poisoned, painting a terrifying portrait of a region where it was, allegedly, quite normal for people to 'drop off short' after being 'white powdered'. They described an uneducated, Godless place, and on 20 September another body was exhumed.

Nathaniel Button had lived and died in Ramsey, a village near Wix, where Mary May had lived with her first husband. Button had been entered in a burial club by his wife, and he died in 1844. It was said she had murdered him to marry again, just like Hannah Southgate – except, by 1848, Button's widow was yet to remarry.

The body was so decomposed that Taylor could not find the stomach in the adipocerous mass in the abdomen. There was nothing for it but to cut up what remained of the organs and perform his usual tests on them. Taylor devoted a week to his tests, and made himself ill as a result: 'I cannot tell you the amount of mental and bodily labour this case has cost me. ... I have almost been poisoned with the effluvia,' he told the coroner in his report, 'and for the last few days have suffered from a desperate headache.'

He did not find any arsenic. Had it been there to begin with, it could have been washed away by water seeping into the coffin in the damp grave, or have been dissolved by the ammonia produced by decomposition. It could, Taylor said, have escaped due to the putrefaction of the organs. But he didn't think Button had been poisoned. Taylor had paid attention to witness reports of Button's symptoms. Even though they were not medical experts, and were relying on memories that were four years old, they did not sound to Taylor's ears like the symptoms of arsenic poisoning. Taylor urged the coroner to compel the jury to return a verdict of natural causes. After half an hour's deliberation, the jury returned an open verdict, in case other evidence against Button's widow came to light.

The pub where the inquest meeting took place was full of locals, peppered with press reporters. The coroner used the opportunity to condemn the rumours that were circulating and that had appeared in the press. He had looked into the cases that had been presented to him, and had decided not to proceed. Perhaps there were some instances where people had been 'unfairly dealt with', and he would investigate them – but only when there were sufficient grounds. He would not be bullied by penny-a-liners who dealt in unfounded gossip.

For the security of society

Appropriately, it was in 1848 that Taylor's book *On Poisons in Relation to Medical Jurisprudence and Medicine* was first published. Poisons occupied much of his professional life, as he travelled the country from pub inquest to courtroom. They had taken up a lot of space in his *Manual of Medical Jurisprudence*, so by devoting a separate book to toxicology, he could treat the subject in far more detail.

Writing from his home in Cambridge Place, Taylor's introduction justified his book. He made his usual argument: poisoning was on the increase, and his book was necessary for 'the security of society'. He had collected and arranged, 'in a convenient form for reference', the important facts relative to poisoning deaths. 'While they constitute a safe guide to the barrister and medical practitioner, [they] may prevent the condemnation of the innocent, and insure the conviction of the guilty.'

Chapter 9

The Formidable Scourge
1849–50

Vapours

Taylor's acquaintance John Snow had been experimenting with ether for pain relief for some years, his articles appearing in the *London Medical Gazette* under Taylor's editorship. When a doctor in Edinburgh announced success using chloroform, Snow changed the focus of his research. From 1848 to 1851, the *Gazette* ran an eighteen-part series where Snow wrote about establishing precise doses for ether and chloroform, and explored the function of other narcotics.

Consulting with Taylor, Snow had come up with tests that could be used to detect the presence of chloroform in human subjects, in breath and bodily fluids. It was an important step, as Snow, being a physician at St James's Infirmary as well as a medical jurisprudence lecturer, knew that chloroform could be used for nefarious purposes. And accidents could happen in hospitals; Snow and Taylor had trialled the tests after a boy had died at Guy's having inhaled chloroform vapour.

Silent cross-examination

In March 1849, Taylor was at Chelmsford again for a poisoning trial. Hannah Southgate had been in prison since August the previous year, yet her feminine charms had not been damaged by the experience. Her barrister, Serjeant-at-Law William Ballantine, described her in his memoirs as 'a young woman of somewhat prepossessing appearance'. The *Ipswich Journal* went to great lengths to describe her outfit: a squirrel tippet, a black veil, and a squirrel muff, 'altogether in the attire of a respectable farmer's wife'. She wasn't a Mary May, scraping by to make a living; she could afford to dress well, and to hire a barrister who was usually to be found at the Old Bailey.

Ballantine demolished the prosecution. The verbal evidence of the village locals was nothing but small-minded gossip, he claimed. But there was no denying Taylor's evidence; Hannah's first husband had died of arsenic poisoning. The prosecution asked Taylor how many grains of arsenic he had found, and what was a lethal dose. Crucially, they neglected to ask if the amount he had found was comparatively large or small. Ballantine did not like scientific or medical evidence; he felt it derailed the legal process. He asked only one question: was Taylor sure he had found sand? Yes, he was. Ballantine didn't ask him to elaborate.

During his closing speech, Ballantine said the evidence wasn't strong enough to support so serious a charge, blaming 'female gossip and chattering'. He did not create the illusion of Hannah being a paragon of virtue, and blamed the poisoning on poor housekeeping. As she had spread arsenic onto bread and placed it around the house to kill rodents, who was to say that she hadn't used a dirty knife when spreading butter on some toast to give to her husband? It was, after all, but a tiny amount.

When the judge, Baron Parke, summed up, he followed Ballantine's lead, and emphasised the small amount of arsenic that had been found. Taylor was sitting on the bench near Parke – the two men had been in court before, at the trial of Hannah Bowyer. Then, too, Parke had been noticeably on the side of the defendant. While the jury were deliberating, Parke 'remarked to him that he was surprised at the amount of arsenic found; upon which Taylor said that if he had been asked the question he should have proved that it indicated, under the circumstances detailed in evidence, that a very large quantity had been taken.' Taylor didn't volunteer evidence, and it was the fault of the prosecution for not putting the question to him. Parke could not use information discovered by accidental means, and Ballantine wrote in his memoirs, 'I had my first lesson in the art of "silent cross-examination".'

The jury returned a verdict of not guilty, and Hannah went to London, where she and her second husband ran an eating house under an assumed name. In her native Essex, the expression 'lucky Hannah Southgate' passed into common parlance. Ballantine, who so hated scientific experts in the courtroom, would meet Professor Taylor again.

Even the centre of the heart

In May, another arsenic poisoning case came to light, this time on the south coast of England. Richard Geering had died in September 1848. The Poor Law Union surgeon, Pocock, recorded the cause of death as heart disease. Then, in December, Geering's 21-year-old son George died. Another son, 26-year-old James, fell ill in February 1849, lingered for some weeks, and finally died in March. Like William Constable in Essex, they had all been in a burial club. They had died of the symptoms of arsenic poisoning, and just before their deaths, their mother Mary Ann had pawned their clothes. She and Richard had often quarrelled about the money he had in a savings bank and just before he died, she withdrew the money without his knowledge.

In April, another son, 19-year-old Benjamin, fell ill with stomach pains, a burning throat, vomiting and diarrhoea, after eating a meal cooked by his mother. Pocock realised they were on the verge of a fourth death in the Geering household, so he called in a colleague, and Benjamin was only fed food prepared by the medical men. He started to recover, so Pocock sent some of Benjamin's vomit to Taylor, and he found arsenic.

The three dead Geerings were exhumed, and their stomachs were sent for Taylor's analysis. He found at least seven grains of arsenic present in Richard, in the stomach, liver, gall bladder, and 'even the centre of the heart'. There was also arsenic present in James, but none found in George, in whom 'a quantity of quicksilver was visible'.

A large, roughly made homemade pill was found in the house, which Taylor's analysis found to contain arsenic and opium. When the case came to trial in August, at the Lewes Assizes, several witnesses came forward to say that Mary Ann had been buying arsenic, even though there were no reports of the Geerings having a rodent problem. Her defence claimed there was no motive for the murders, but the jury, who deliberated for only ten minutes, did not agree. They found her guilty, and she was sentenced to hang.

The huge amount of arsenic found inside Richard Geering mitigated against an accidental poisoning. If Mary Ann was guilty, had she been inspired in her plan by the coverage that burial club murders had received in the press, and had she thought that she could get away with it? She had received about £15 in total from the burial clubs, along with a pittance from the savings bank; was that enough money to have murdered so many of her family?

Forensic testing proved its worth in this case. Because Pocock had sought Taylor's intervention, they had saved the life of Benjamin Geering. There was, however, the issue of George Geering. His mother might have adjusted her *modus operandi*, and used mercury to kill him instead of arsenic. But he might not have been murdered at all, instead being accidentally overdosed by Pocock, who was fond of dispensing mercury pills.

Before I go hence

In July 1849, an elderly lady died on Albany Street, very near the Taylors' home. Mary Cancellor, the aunt of Taylor's wife, Caroline, was the last surviving sibling of Caroline's father. A spinster, Mary had possibly helped in the upbringing of Caroline and her brothers after their mother's death. That she lived so near to the Taylors suggests a continued closeness, which is obvious from her will. She left Taylor £10, and amongst other legacies was £50 for the Taylors' daughter. Caroline was left a sable tippet and half of the rest and residue of her aunt's estate.

Mary Walker, the Taylors' housekeeper, was also remembered in the will. As she was in her seventies, it's possible that some years before, she had worked for the Cancellors.

One of the witnesses to the will was Taylor's friend, the surgeon John George French. He may have taken Mary on as a private patient, but the fact that he witnessed the will of Taylor's relative shows the continued close acquaintance between the two men.

The omnipotence of stenches

In the summer of 1849, cholera arrived in London. Fear of it had panicked the populace for some years, and it had made landfall in Edinburgh at the end of 1848. Cholera haunted the *London Medical Gazette*, with articles and letters appearing beside the journal's usual fare of lectures and surgical tips. Helpless, each week, all Taylor could do was to include the number of deaths caused by the 'formidable scourge' in his editorials, while he complained about related issues. Burials in London churchyards came under attack from his angry pen, as did the membership of the General Board of Health.

He criticised board members Dr Southwood Smith and Edwin Chadwick for their enthusiastic 'faith in the omnipotence of stenches', which meant that their views on 'the subject of contagious and epidemic diseases cannot be received with respect by the profession'. He had even less time for the two lords on the board, whom he criticised for not being qualified. They seized 'upon every scrap of government patronage which can be obtained', blocking the way for trained medical professionals who 'labour year after year, through penury and adversity, battling with all the ills of life' to 'do good in their day and generation – to make mankind healthier, happier and better – and this often without hope of reward'. His readers would have well understood his frustration.

Deaths from cholera rose steadily in London from early June, with twenty-two in the first week, almost doubling to forty-two in the second week. The number increased only slightly to forty-nine in the third week, but leapt up in the fourth week to 124 deaths, and 152 in the fifth. By 20 July, the number had more than doubled, to 339. Worst hit were areas on the south of the Thames. 'About one-third of the fatal cases occurred in the districts of Lambeth, Rotherhithe, and Bermondsey.' This was where Guy's Hospital was located. The epidemic worsened; the weekly statistics showed that by the middle of August, an average of 176 people a day were dying of cholera in London, and complaints were made in newspapers that the London hospitals weren't taking in cholera victims. Taylor protested that Guy's had set aside some outbuildings to accommodate them, but they could only take in a limited number. Besides, 'the lives of some hundreds of inpatients afflicted with other diseases are at stake.'

Articles and letters by medical men tried to explain what cholera was – causes, treatments, preventions. Some sent diagrams of 'material' that had been brought up or 'evacuated' by cholera patients, viewed through microscopes. John Snow had a theory: he wrote a pamphlet suggesting that cholera was waterborne. He had hit on it exactly, but his theory was not accepted. Even Taylor, who did not believe that miasma theory explained how cholera spread, thought Snow's theory was novel, but didn't see it as an 'unquestionable truth'.

Taylor wrote editorials about the suggested cures for cholera, his sarcasm desperate in the face of enormous human tragedy. There were opium lozenges recommended as a preventative. It was possible, Taylor wrote, that 'the free exhibition of them to children may add to the number of coroner's inquests.' Two medical gentlemen had suggested wrapping patients in wet sheets, in order to restore heat to the surface of the body. Taylor saw hydropathy as quackery, and began his dismantling of their treatment by joking, 'We by no means wish to *damp* the ardour of hydropathists …' Another doctor had suggested freezing parts of the alimentary canal, believing that cooling the body was a cure; Taylor said he would leave the gentlemen to fight it out between themselves. A Spanish doctor recommended taking large amounts of olive oil, a treatment that Taylor believed 'baffles all theory', and a British doctor suggested swallowing salt at every meal 'in heroic doses'; his theory was that the salt would release chlorine, which would disinfect the stomach. Taylor did not agree.

Squeezed in between all this was a report of the Anniversary Meeting of the Provincial Medical and Surgical Association. A paper on 'secret poisoning' had been read, which echoed Taylor's own concerns; it was a crime 'which had prevailed of late years to a most fearful extent'. Some recommendations were put forward: that no druggist should sell arsenic without a license; that no one should be able to buy it unless accompanied by a witness, and unless they can show why they need it; and that the vendor should keep a book, entering every sale, including the purchaser's name and address. They believed this would aid detection in cases of murder, as well as put murderers off. In the articles and letters on the epidemic raging at the time, the disease was often described as 'the cholera poison'; whilst cholera was invisible, its cause mysterious and treatments apparently doomed to fail, that other killer poison, arsenic, could be controlled.

At the height of the cholera epidemic, William Odling, 19-year-old son of police surgeon George Odling, went to visit Taylor at Guy's with a stomach to examine. A famous stomach, it had belonged to Patrick O'Connor, the alleged victim of husband and wife Frederick and Maria Manning. Amongst all the horror of the cholera epidemic, it was one death that garnered a huge number of newspaper columns and was the sensational crime of the day.

O'Connor had disappeared on 9 August. His body was found eight days later buried in quicklime under the floor of a kitchen in Bermondsey. It was clear from the wounds to his head that he had died after being shot and repeatedly beaten about the skull by a blunt object. A half-empty bottle of laudanum was found in the house by police so, as a formality, his stomach needed to be analysed.

When William Odling arrived at Guy's, Taylor wasn't there. The professor would not have had teaching duties in August, and with Guy's being in one of the worst hit areas of the epidemic, it's no surprise that William found him absent. Whether the stomach was left at the hospital by Odling, or if he took it back to his father isn't clear, but Taylor refused to touch it. He wrote a letter to George Odling, in which he 'informed him that his reason for so declining arose from the fact that the counties of Surrey and Middlesex never paid him for his trouble in such matters.'

It was one of the most famous murders of the nineteenth century. O'Connor had once been Maria's lover, and he was known to have some money stashed away. He was last seen dining with the Mannings on the day he vanished. The Mannings had taken his money and run, but in different directions. Maria had gone to Scotland, and Frederick to the Channel Islands. Trains and telegraphs had been used once again to catch murderers, and husband and wife were arrested. Maria gave the proceedings some glamour; she was Swiss, and had been a servant to various aristocrats. It was a crime of sex and death, in the midst of a terrifying epidemic. And Taylor dug in his heels.

It wasn't unusual for Taylor's letters to coroners to be printed in newspapers, so he might have hoped that his complaint would be published. He had had enough of not being paid for his gruelling, time-consuming work, which sometimes took a toll on his health. After making his laboratory analyses, appearances at inquests and at trial would usually follow, increasing the amount of his time taken up by each case.

'A month seldom passes without our receiving one or more communications from medical practitioners regarding the ill-treatment which they have sustained at the hand of coroners,' Taylor wrote in the *London Medical Gazette*. Letters often appeared in its pages from angered surgeons who had not been paid. Taylor would have seen the huge press coverage that the O'Connor murder was receiving and perhaps realised that if he refused to work on this case, his grievances would reach a wide audience.

It was clear that opium was not the cause of death. The fractures caused by the blows, George Odling told the inquest, were 'sufficient to cause death half a dozen times over, were it possible'. Besides, O'Connor had been dead for over a week by the time his body was found, and from experience, Taylor would have known that after so long, any opium would have been almost impossible to trace.

George Odling did not feel competent enough to perform the analysis himself. He did not approach another analyst, perhaps assuming they would have refused as well. The organ was useless to him, and Odling destroyed it. As cholera was known to be associated with the stomach, it is perhaps no surprise that Odling wanted to be rid of it.

Had Taylor performed the analysis, he would have shared a courtroom again with Ballantine, who defended Maria Manning. Husband and wife were found guilty, and were executed side by side outside Horsemonger Lane Gaol by Calcraft; amongst the enormous crowd was Charles Dickens. Maria 'was the power that really originated the deed of blood,' Ballantine claimed in his memoirs. With the Mannings, the cholera epidemic died away, and a new decade dawned.

I thought the prisoner was joking

On 25 February 1850, another stomach found its way to Taylor's laboratory. It had belonged to Susan Lucas, who had died three days earlier at her home in Castle Camps, Cambridgeshire. She was in her twenties, and had been married to Elias Lucas for four years. Susan's sister, 20-year-old Mary Reeder, had moved into the couple's home in January after Susan had given birth to a child who had not survived. By February, Susan was in her usual health again. The day before she died, she had been seen in the local shop buying ordinary, everyday items: bread and a pair of socks for her three-year-old.

Susan fell ill with vomiting that evening, and the next day, her husband went for a surgeon. By the time he arrived, Susan had been dead for half an hour. 'I observed that she died in a state of collapse,' the surgeon said, 'the fingers had a bird's claw appearance, and blue, as though she had died of cholera; I felt her pulse, and said I was sorry I had not been called in before.' Her stomach was blue as well, which again suggested cholera, but there was an almost entire lack of diarrhoea, so the surgeon demanded 'a further investigation into the case'.

At this point, Mary began to crumble. She said, 'To tell you the truth, my sister has been much worse since the water-mess last night, and we are of the opinion that something has produced her death.'

At inquest and trial, locals came forward to give evidence; things they had overheard, or conversations they had had with Mary and Elias, which put a suspicious complexion on Susan's sudden death. One claimed that Elias had said, 'I wish I could get rid of my wife, for I have a bastard child coming.' The witness commented, 'I thought the prisoner was joking.' After his wife had given birth, he told the midwife he'd wished he'd never married Susan. She cautioned him not to tell his wife such a thing at so difficult a time; yet he did so, and the midwife found the exhausted woman in tears.

A set-to between Susan and Elias was recalled, where he had remarked that he was saving one of the pigs until he was married to her sister.

'That will never be, for they will never allow you to have my sister.'

'They can't help themselves if I go a little away from home.'

At the time, a marriage between a man and the sister of his deceased wife fell within the prohibited degrees, Biblical teaching looking on it as incestuous. The marriage could take place, but it could be open to legal challenge. The local vicar could refuse to marry the couple if he felt strongly enough, so devious Elias would slip away to marry.

Taylor had been sent Susan's viscera and, unusually, some earth from the Lucas's garden. At the trial in Cambridge, only a month after Susan's death, Taylor explained that he had found two grains of arsenic in Susan's viscera. He did not find any ulceration, but her death had been very quick. She appeared to have been given a vast amount of arsenic, more than the two grains he had found, as there was inflammation throughout her stomach and intestines.

The evidence of the earth from the garden was barely mentioned in the press, but Taylor wrote about it in the 1859 edition of *On Poisons*. Before Susan had taken to her bed, she had gone outside to be sick. By testing the earth, it meant, in effect, that her vomit could be tested. 'I found arsenic in a few ounces of this earth, in a soluble form, and in rather large proportion.' To ensure that the earth in the garden wasn't naturally riddled with arsenic, other soil from the garden was examined. 'In earth taken from the path at a distance from this spot, no arsenic was found.' It isn't clear if it was the idea of the police to send the earth to Taylor, or if something they had told him prompted Taylor to have them send it for his analysis.

In the courtroom, Taylor was passed some powder that had been found in the Lucas's home. Was it arsenic? If he was allowed half an hour to test it, Taylor could find out. It raises an intriguing question: when Taylor went to inquests and trials, was it usual for him to bring his chemical-testing equipment with him? At the trial of Mary North, he had borrowed a local chemist's laboratory, but that had involved oil of vitriol; this case involved arsenic, and the Reinsch test required simple apparatus. Taylor may have heard that a packet of what was thought to be arsenic had been found in the house, so he had come prepared.

He returned to the courtroom and informed the assembled crowd that it was arsenic. The jury found Elias and Mary guilty, and they were sentenced to hang. As they awaited their meeting with William Calcraft, both confessed to the crime, Mary claiming that Elias had encouraged her to kill her sister.

It made his teeth shake

In April, the same month in which Elias and Mary breathed their last, Taylor examined some poisoned coffee. Louisa Susan Hartley was 18. She had been in service, but ill health prevented her from working, so she had come home to live with her widowed father, Joseph, in one room of the Hope public house in Southwark. Joseph had to support her financially, struggling as an out-of-work Fellowship Porter on the Thames. There was evidence that he had beaten Louisa with a stick, and she had shown the bruises to a neighbour, who reported that Louisa had laughed and said she would poison him if he did it again.

Louisa usually made her father's breakfast; one morning that April, she had taken him coffee and some bread and butter.

Joseph claimed to have been asleep, and hadn't seen her prepare breakfast. He tasted the coffee, and spat it out because it had an unpleasant taste. He asked Louisa if she had put something in it. 'No, father,' she said. He tried it again, and once more spat it out. 'He felt a parched and burning sensation in his mouth' and 'it made his teeth shake.' He accused his daughter of putting poison in the coffee, and Louisa took the cup and threw the contents in the washbasin.

He watched his daughter wash out the cup, then he tasted the coffee she had poured out for herself; he 'did not perceive anything the matter with it' and he accused her again. He put the contents of the basin into a jug and took it to a doctor, then on to nearby Guy's Hospital. When he got back, Louisa was crying, and Joseph 'asked her if she did not think it was a very bad job to be trying to poison him'. She didn't reply, and Joseph went off to find a policeman to arrest her. When the policeman arrived, Louisa told him that she had accidentally dropped some soda into her father's coffee. Along with the contents of the basin, Joseph gave the police an array of items from their one-room home: 'the cup in which his coffee had been; the coffee-pot; a piece of soap; and a bottle'. These found their way into Taylor's hands.

Taylor found that the coffee had oil of vitriol in it, which explained Joseph's symptoms. He wanted to find out when the acid had been added – either in the pot, or in the cup after it had been poured. If the poison had been added after it had been poured, it might suggest that Louisa was innocent. Taylor believed that 'this question was solved by the aid of chemistry'.

He observed that the coffee pot was old and rusty. He tested the coffee from the basin and found that it didn't contain any iron. He warmed a small quantity of the poisoned coffee in the pot, and 'it was immediately and strongly impregnated with sulphate of iron.' Taylor claimed in *On Poisons* that this clearly showed that 'the acid had not been mixed with the coffee in the pot, and might have afterwards been put into the cup' without Louisa's knowledge. Any acid in the coffee would have reacted with the rust in the pot; it hadn't until Taylor performed the test, so it proved that the poison was added to the cup after it had been poured out.

The case went to trial at the Old Bailey in May. Louisa was defended by Ballantine, who for once would have been glad of Taylor's evidence. In court, Taylor explained how the amount of acid he had found was tiny, insufficient to cause death, and would not have caused permanent injury. It weakened the case against Louisa, as the evidence was starting to suggest that Joseph had poisoned the coffee himself.

Ballantine's closing speech emphasised that it was 'a dreadful thing to see a young girl' standing at the bar on 'a charge of attempting to poison her own father', and that it was 'equally dreadful to see a father pressing a charge with the determination and vindictiveness' that Joseph had. Ballantine said that a man who beat his daughter as Joseph had 'was devoid of all proper manly feeling' and he suggested to the court that Joseph had falsely accused her 'for the purpose of getting rid of his daughter, or from some other equally bad motive'. The judge summed up in Louisa's favour, finding Joseph's testimony suspicious. He had got 'it up in the most cool and systematic manner, and [had] not let any single point escape attention which was likely to lead to a conviction of the prisoner'. Louisa was acquitted by the jury.

To ensure Louisa's future welfare and safety from her brutish father, she 'would be taken under the charge of the Ladies' Committee of the gaol'. But in March the following year, she was still living in Gravel Lane with her father. A few months later, he married a straw bonnet maker twenty-four years his junior, who evidently did not mind that her new husband had got up a false poisoning charge against his own child.

In such small quantities

Five days after the Hartley trial, on 16 May 1850 a man in a small Essex village died. He had been suffering from tuberculosis for several months, but because his widow was Sarah Chesham, an inquest was opened into his death. Hawkes, the surgeon who had attended him in his illness, was the same man who had ministered to Sarah's sons as they died of arsenic poisoning in 1845. He had been roundly criticised for registering one death as 'English cholera' and entering no cause of death at all for the other child. Going back into the Chesham household, surely Hawkes would have been on his guard, looking for any signs of poisoning again?

As he gave evidence at the inquest, he said that he saw nothing untoward: Richard Chesham 'was apparently dying with consumptive symptoms' and the post-mortem showed that 'the lungs were ulcerated and full of tubercules'. Despite his testimony, the jury wanted the assurance of Taylor's word, so Richard's viscera were sent to him. At the same time, one of Sarah's surviving sons was ill; it would have been sensible, as with the case of Mary Ann Geering, to have analysed anything he vomited, but this does not appear to have been done.

At the end of May, Taylor sent news to the coroner: he had found arsenic in Richard Chesham's remains. Not, as he told the reconvened inquest, a vast amount. It was 'in such small quantities as to prevent him stating positively that the poison was the cause of death', and in the whole of the viscera sent to him, he had found only one-twenty-fifth of a grain. But an item removed from the Chesham's cottage – a bag of rice bought by Sarah's elderly father – contained huge amounts of arsenic, twelve to sixteen grains, which was enough to kill six adults. When Taylor looked in the bag, each grain of rice was coated in white powder, which, when tested, proved to be arsenic. At the inquest, Sarah had sworn that she hadn't fed her husband rice despite other witnesses saying she had; here was evidence that surely proved how she had administered the arsenic.

If the quantity was so small, how could Sarah be the murderer? Because, Taylor said, she had slow-poisoned an already dying man. 'Small doses would have the effect of accelerating death – they would excite great pain and anguish, and debilitate and exhaust the patient; and where there were consumptive symptoms they would tend to weaken the bodily power.' Despite Taylor's confident testimony, the jury remained unsure. They deliberated for half an hour, and returned their verdict that Richard had died of consumption, the evidence not showing that his death had been accelerated by arsenic. Sarah Chesham was free, but the coroner, Charles Carne Lewis, decided to refer the case to the local magistrates.

During the summer, a secret investigation, hidden from the press, went on into the death of Richard Chesham. They wanted to convict Sarah for poisoning with intent to murder. The investigation was carried out by Lewis, the Essex police, and magistrates. The authorities were determined to catch Sarah. She was suspected, as was Thomas Newport, who back in 1846 had been under suspicion of having paid Sarah to poison his illegitimate son. Hannah Phillips, a local gossip, clacked her tongue. She claimed that Sarah had admitted to poisoning Lydia Taylor's baby, and that Sarah had offered to poison, or 'season', a pie for her, to kill Hannah's husband. She reported that Sarah told her she had at one point hidden arsenic in a tree stump.

The authorities asked Taylor for his opinion. He didn't think it at all likely that Sarah could be convicted. It stood 'too much on presumption' and 'there is a want of that strong medical proof which is necessary for a conviction.' He did, however, think that Richard's bouts of illness had been caused by the administration of small amounts of arsenic. The advice of a barrister at the Middle Temple was sought, and he too thought the case was on shaky ground, but he put forward suggestions on how they might proceed. He even directed that Taylor should examine the tree stump for signs of arsenic. There is no evidence that Taylor ever did; he must have sighed at the very idea of it. Testing soil that someone had recently vomited on was one thing, but if rainwater could wash arsenic out of a grave in a soggy churchyard, then what possible chance did it have in a tree stump?

Strong medical proof or not, in early September, Sarah Chesham was arrested for a second time.

The diabolical art

William Bristow, aged 34, had lived in the Lincolnshire village of Quadring. He died suddenly at the beginning of October 1850, his symptoms being those 'of irritant poisoning by small and long continued doses', and his widow was the suspect – an opinion that might have been influenced by the new twist in Sarah Chesham's story. The *Lincolnshire Chronicle* ramped up the drama for their readers; slow poisoning was 'a mode of despatching persons in great repute in the middle ages; a Marchioness of Brinvilliers being a prominent character, and proficient in the diabolical art, as was Madame Lafarge, in our own times.'

William's widow had suddenly lost another husband before him, and a child. Had she poisoned them too? Taylor's opinion was required. But it does not seem that he went to the resumed inquest meeting to give his report; there really was no point. 'He was unable to discover any traces of poison whatever,' and gave his opinion that death was caused by the simultaneous inflammation of stomach and intestines – gastroenteritis.

The *Chronicle* marvelled at Taylor's results. 'It is a happy thing that recourse was had to Dr Taylor, as the result has removed all suspicions against the unfortunate wife, and cleared her character from an imputation which, had a less rigid inquiry been instituted, would have clung to her for life.' Or would even have ended her life. There was no Marchioness de Brinvilliers in Quadring, no Madame Lafarge at large in Lincolnshire: Mrs Bristow was free to go.

Chapter 10

His Very High Position
1850–54

The belle of the village

On the evening of Monday, 14 October 1850, Superintendent Thomas Coulson of the Essex Constabulary arrived at Taylor's home. He brought with him a pair of stained corduroy trousers, and Coulson needed to know if the marks were blood.

The trousers belonged to 23-year-old farmer Thomas Drory, from the village of Doddinghurst. Drory's father, James, was well respected in the area. On noticing that Thomas had developed an 'improper intimacy' with Jael Denny, the stepdaughter of James's farm bailiff, he gave the bailiff notice to quit. But after James left the farm to Thomas's management, his son took up his intimacy with Jael once again. She was only just 20, unusually tall at 5 feet 9 inches, 'and was generally considered the belle of the village.' Thomas was 3 inches shorter than her, had a 'youthful appearance and good looks', with 'fair hair, a brilliant florid complexion, small delicate features, regular in their contour, and a general cast of countenance very expressive of mildness and rustic innocence'. When a story caught the public attention, the details compiled by the newspapers were novelistic.

Jael had fallen pregnant. No matter what a 'belle' she might be, Thomas was from a different class and would not marry her; he wanted a farmer's daughter for a wife. In August, he had got hold of some poison for Jael to induce a miscarriage, but it evidently had no effect; by October, she was about to give birth. She was last seen by her mother on 12 October. She said she had arranged to meet Thomas at half past six that evening. 'I will finish my tea when I come back; I shall not be gone long, I am only going to the first stile.' She put on her bonnet and cloak, and was never seen alive by her parents again.

They waited up all night for Jael. Remembering the incident with the poison, they were worried that Thomas had harmed her. At daybreak, her parents went out looking for her. After two or three hours, her stepfather spotted 'what he thought to be an ox lying on the grass in a secluded part of the meadow, which is overshaded by a thick clump of trees'. As he got closer, he saw that it was the body of his missing stepdaughter. He cried out, loud enough that people from nearby farms came running. They turned over the body and saw 'that she had been strangled by a rope which had been twisted several times round her neck'. There was blood on her face and her clothes, her features 'shockingly distorted'. The rope had cut into her neck, and there were marks on her hands 'as if they had been bitten or torn', indicating 'the desperate struggle she had been engaged in, and the utter impossibility of her having committed the act herself'. Her bonnet was bent in several places, and her cloak was ripped. She had met her end at violent hands.

Information reached Superintendent Coulson, stationed 4 miles away at Brentwood. He set off to trace Thomas Drory, and found him lounging at home by the fire. Coulson insisted he come with him to view the body, and Drory followed, but hesitated as they reached the meadow. On seeing Jael's body, 'Drory became deadly pale, and could scarcely walk.' Coulson arrested him.

A remarkable number of forensic clues had been identified by police and by the surgeon who had carried out the post-mortem. There was her clothing, which had been damaged during a struggle. There was the cord that had been used to strangle her; police had found an identical piece at Drory's house, the end of which had been recently cut, and matched up exactly with the length that had killed her. The coroner, Charles Carne Lewis once again, later said that this could not be established, but the fact that the police had thought to search for the rope says much for how investigative methods were evolving.

The medical evidence showed that Jael had suffered a violent attack. The rope had been slipped around her neck from behind, which meant that her attacker could have been shorter in stature than her. Scratches and bite marks on her arms and hands showed how desperately she had fought for her life. There was a bruise on her chest, which the surgeon suggested had been caused 'by a person kneeling upon her chest before death'.

What was needed was evidence to link Thomas Drory to the scene. Jael's parents knew she was going to meet him, and a local had seen Jael 'in Drory's company, walking over some meadows'. But was it blood on his trousers?

The inquest was held at the courtroom in Brentwood's police station, packed with curious locals. One of Jael's sisters sobbed hysterically. When Thomas Drory was brought into the room, he shook hands with his father and his uncle with calm self-possession.

Taylor gave evidence, at what appears to be the first case at which he commented on bloodstain analysis. His report, which was apparently lent to the press, as it was printed at length in several newspapers, described in detail the size and exact location of nine different stains on Drory's trousers, the appearance of each stain, and how far it had leaked through the fabric. He had used a 'powerful magnifier' and with some of the stains, he had found 'small dark red coagula, resembling hardened or clotted blood, and presenting the glistening and shining character of dried and clotted blood'. Two large stains had the appearance of having 'been wetted and wiped' and that 'the colour was unlike any stain except that arising from blood.'

Nearly every stain was blood, except one, 'a small red stain, as if from fruit'; it being October, there were all manner of wild berries in the countryside. Taylor explained that 'a great many substances will give a red colour to clothing' without being blood, but that none, except blood, would cause a clot. For a clot to form, it would have to be from 'a living person, or one recently dead'. He had asked 'two surgical friends' to examine the stains, and they concurred with Taylor. It's possible that he had consulted George Owen Rees, his colleague from Guy's, who was an expert on blood.

Great violence must have been used for so much blood to have been produced from strangulation, far too much for it to have been a suicide. It ruled out an argument that Drory had been beside Jael as she strangled herself. Despite Drory trying to convince people that Jael was suicidal, the only way for the blood to get onto his trousers was if he had pulled the cords.

There was no way to prove who the blood had belonged to. A juryman asked a question, phrased in a Victorian manner: 'Is there any difference in the blood of an animal which a farmer is apt to get on his clothes, and that of a Christian?' Taylor had to reply that there wasn't. But Drory had told Coulson that the stains were from food he was feeding the cows, and had presented the policeman with a pot, full of cobwebs and evidently long out of use. Even though Taylor couldn't prove the blood was Jael's, nor even if it was human, he had been able to prove that Drory had lied.

Taylor's opinion was asked on another problem. Jael's body was still warm when it was found, yet she hadn't been seen since the evening before. There was evidence that Drory had gone to Brentwood on the morning she was found, so could the temperature of her body be evidence that he hadn't killed her? The issue of the time taken for bodies to cool after death was a very inexact science in 1850, but Taylor was able to make observations. He said he had known one case where the trunk was still warm after thirty or forty hours. He thought it was 'quite possible that a woman so advanced in pregnancy, and killed at half-past six o'clock, might be found warm at ten next morning, but a good deal depends on the clothing.'

Taylor's evidence, along with that of Doddinghurst locals, combined into a strong case against Thomas Drory. The inquest jury returned a verdict of wilful murder, and Superintendent Coulson took the suspect to Chelmsford gaol to await trial.

Jael was buried in the churchyard at Doddinghurst. The funeral was conducted privately, although from the number of newspapers that reported it, several journalists must have lurked in the lanes for a view; *The Chelmsford Chronicle* even reported how deep her grave was. The murder had attracted the horrified attention of the public. Scores 'went to view the body of the poor girl, which a correspondent says was exposed, he considers indecently, even up to the morning of the funeral.'

It was discovered that 'some persons fond of treasuring the horrible … had cut up and carried away the turf on which the death struggle took place, and which was stained with the unfortunate woman's blood'.

Her stepfather fell ill after the funeral and wasn't able to work. Whereas some people had ghoulishly run off with blood-soaked turf, others made financial donations to Jael's family. They came from all over the country, from all walks of life. There was Admiral Sir George Mundy and Lady Kemeys, various reverends, and from Portsmouth, a servant called Eliza Fowles. With her donation, she wrote, 'Will you please give the enclosed small trifle to those poor afflicted people? I am sorry it's not more.'

'Lafarged'

Taylor's celebrity was spreading; he was mentioned in an article that appeared in Charles Dickens's magazine *Household Words* on 23 October. The writer was in France, where he met various people, including a charming widow. He panicked: perhaps the widow was 'familiar with drugs and hallowed potions'. Had her husband 'been so to speak, Lafarged …? I should like to hold inquest on remains of deceased … to have him exhumed, and the contents of [his] stomach put in a jar and analysed by Professor Taylor.'

The name 'Professor Taylor' had become as associated in the popular mind with poisons as that of the notorious Madame Lafarge. But whereas she was the perpetrator of terrifying secret crimes, Taylor was the indomitable nemesis of those who followed in her wake.

A new poison

William Wren was a very naughty boy. Aged 14, he had been convicted of sheep stealing, and served eighteen months in prison. After his release, he took a job working for Thomas Claridge, a market gardener in Millbrook, near Southampton. Aged 16, he was accused of trying to poison Claridge's family.

On Boxing Day 1850, he had been sent by Thomas's wife Mary to fetch some milk, which she was going to use to make a pudding. It looked dirty, and Mary sent him away. He returned a few minutes later, and Mary said, 'Bill, you've brought me the same milk back, only it's strained.' She tasted it and found it had 'a bitter and acid taste'. She spat it out at once, and exclaimed, 'Oh, Bill, this is poison.' She threw up twice and her tongue felt as if it was burning. Her husband tasted the milk too, and it made him feel ill for two or three hours.

Was it poisoned by the woman who sold the milk, or was it Bill? One of the parish constables was called in, and Groves, the local schoolmaster, said he would go with Bill and the constable to change the milk. As they crossed a stream, the two men saw Bill take a piece of paper out of his pocket and drop it into the water. Groves jumped over the bridge and picked up the paper. It said, in the local chemist's handwriting, 'Nux Vomica – Poison'.

The local chemist sold it for people to kill mice. 'He gave it to persons who asked for arsenic for that purpose as he would not sell arsenic.' It was usually sold as a greyish-brown powder. As its name suggests, it induces nausea and vomiting, which is perhaps why the chemist thought it was a safe substitute for arsenic: it would be difficult to take a fatal dose without throwing up. Nux vomica is one of the seeds that contains strychnine, a poisonous alkaloid that affects the spine, causing agonising, tetanus-like cramp. The body bends backwards, head to heels like a bow, and the victim finally dies from exhaustion or suffocation. Strychnine had only been discovered in 1818, so was a very new poison in 1850.

The inquest opened on 28 December, and the sediment found in the bottom of the milk jug was sent to Taylor to examine. He found twenty-seven grains of nux vomica. His test, using strong nitric acid, should have made it turn a deep orange-red, but instead, it turned deep green. The mill used by the chemist to grind nux vomica had previously been used to grind guaiacum, a resin that had various medicinal uses. Although the guaiacum had interfered with Taylor's analysis, 'it served to identify the sample'.

Wren was tried at the Hampshire Lent Assizes in Winchester in 1851. He was found guilty of attempting to poison, and was sentenced to fifteen years' transportation. But whatever motive he had for poisoning his employer's family apparently never came to light.

No safety for mankind

That same month, two more cases that Taylor had been involved in went to trial; Sarah Chesham and Thomas Drory were in the dock at the Essex Lent Assizes. Perhaps because Sarah wasn't indicted for murder, but the same charge as William Wren, where the severest punishment was imprisonment or transportation, she didn't have a defence counsel. She was near-illiterate and had spent months in prison, so had no hope of getting up a defence for herself, and had not prepared any witnesses to speak on her behalf. Charles Chadwick Jones, who had defended her before, and had unsuccessfully defended Mary May, was bankrupt and in extremely poor health, so he could not help her. In the newspapers, she was overshadowed by Drory and his far more thrilling and horrible crime.

Despite the doubts leading up to Sarah's arrest, the judge and the press decided that the evidence against her was overwhelming. Without anyone to cross-examine the witnesses, there was no one to ask Taylor if the appearances he had seen in

Richard Chesham's viscera could have been caused by tuberculosis, rather than arsenic poisoning. Replying to questions from the prosecution, Taylor emphasised the idea of slow-poisoning, it being the only possible way to explain how the tiny amount of arsenic he had found could have led to death.

Hawkes, the surgeon who had carried out Richard's post-mortem, changed his testimony from that given at the inquest. Originally, he had said that he saw no signs to show that Richard had died of anything but consumption. But now, he said that the symptoms were those caused by 'the administration of small doses of some irritant poison'. There was no barrister defending Sarah who could have demanded to know why he had changed his mind.

The trial was flawed, but the evidence of the arsenic in the bag of rice was incriminating. The only way to explain its presence, unless Sarah had put it there, was that it was an accident, or had been added by someone hoping to snare her. Sarah was found guilty of attempting to poison, and the sentence should have been imprisonment or transportation. Except that the judge, Lord Campbell, sentenced her to hang. He had decided that she was guilty of the crimes she had been acquitted of in 1847, and he was convinced of her guilt now; if he didn't hang her, 'there would be no safety for mankind'. Not that Campbell found it easy to pass sentence. *The Essex Standard* reported that he was 'so affected that he could barely proceed'.

Drory was also found guilty, and he was hanged beside Sarah Chesham. As was common at the time, their crimes were turned into ballads, which were sold by pedlars in the crowd at the execution, and by street hawkers around the country. For sale was an illustration of Drory strangling Jael, which contained a grimly accurate depiction based on the forensic findings, the rope wrapped twice around her neck, his knee pressed onto her chest. [Plate 14]

The day before the execution, a Bill to regulate the sale of arsenic had its final reading in the House of Lords before being sent to the House of Commons. It was passed in June that year. It required anyone buying arsenic to sign or mark a register, in the presence of a witness, just as the medical professionals at the Provincial Medical and Surgical Association meeting had recommended two years earlier. It was to be coloured with indigo or soot, unless it was required for the arts or manufacturing. It had taken time and argument to reach this point; arsenic's usefulness meant its regulation was unpopular with Victorians who thought that nothing should stand in the way of commerce and enterprise.

At home

The census was taken on the night of 30 March 1851. Taylor was listed as a physician. He had become a licentiate of the Royal College of Physicians in 1849, even though he didn't have an MD. There was no mention of him being a lecturer as well; in 1851, his colleague Arthur Aikin retired from Guy's so Taylor had become sole lecturer of chemistry, as well as medical jurisprudence.

Edith, the Taylors' only surviving child, was described as a 'scholar at home'. There is no governess on the census, so Edith might have been taught by her parents. Four of their five female servants had the surname Walker, and they had all been born in the same village in Herefordshire; they were presumably relatives of each other. Mary Walker, named in the will of Caroline's aunt, was the Taylors' housekeeper, but aged 84 was probably not capable of much work. That she was kept on as staff suggests there was a sentimental, rather than practical, reason for her continued employment.

Bearing in mind how often Taylor had investigated mysterious deaths in Essex, it is ironic that his brother Silas had moved to the county. He had made his home in the village of Upminster, which was later swallowed up by Greater London. In 1851 he lived in a large 'French style' house called Fox Hall, which still stands today.

Alfred Swaine Taylor, MD

Scotland had led the way in the teaching of medical jurisprudence in Britain, and in his early days Taylor had much admired Robert Christison in Edinburgh. In January 1852, Taylor was conferred an honorary degree of Doctor of Medicine by the University of St Andrews. It was unanimously agreed, 'in consideration of his very high position as a Toxicolgist and Medical Jurist generally, and of the benefit he has conferred upon Medical Science by his works on *Medical Jurisprudence* and *Poisons*'.

It was a great honour, showing in what high regard his work was held. Taylor's *Manual of Medical Jurisprudence* went into its fourth British edition that same year; in 1853, it went into its third American edition. It had first been published less than a decade before, but Taylor was keen to constantly update it as medical knowledge advanced, and as more cases could be added. He had many correspondents, some of whom were former pupils, telling him about cases they had worked on, and he compiled them carefully into his text.

You are not going to poison me

In North Ockendon, Essex, only a few miles from Silas Taylor's home, a suspected poisoning was uncovered at the end of 1852. Agricultural labourer James Lister had married a widow, Sarah Skingsley. She had had seven children by her first husband, and she had two more by James. After she died in April 1850, her daughter Ann gave birth to two children. The 1851 census shows James, aged 38, with his stepchildren, 21-year-old Ann and 13-year-old Richard Skinglsley; Henry Lister,

6, his son by Sarah; and one-month-old Alfred Lister. A note says that Alfred is 'the son of the above Ann Skingsley', and it appears that Alfred's surname at birth was registered as Skingsley, not Lister. It was assumed that James and Ann were living 'in concupiscence' – their relationship was within the prohibited degrees of the church. They had been living rent-free in a former parish workhouse building, but when it was suspected that Ann's children were by her stepfather, they were evicted. James and Ann often quarrelled, and Ann had been heard to say, 'You are not going to poison me as you did my mother.'

North Ockendon's vicar intervened, and he and several other locals forwarded their suspicions to coroner Charles Carne Lewis. Lewis had Sarah Lister exhumed; her body was nearly black and partly decomposed. Superintendent Coulson, who had worked on the Drory case with Lewis, took the viscera to Taylor. The case was widely reported, the relationship between James and Ann adding a prurient angle to the suspected poisoning. Taylor found that Sarah Lister had not been poisoned, but this humdrum outcome wasn't as widely reported as the lurid facts that were gleefully circulated before.

Persecuted and slandered to the death

Taylor's repute was such that in 1853, a Dublin solicitor staked the freedom of William Bourke Kirwan on his word. Kirwan was a miniaturist and anatomical artist, with a wife who enjoyed swimming. They had gone on holiday to Howth in September 1852, making frequent trips to the rocky island of Ireland's Eye, where Kirwan would sketch and paint, while his wife Maria went swimming. One evening, when the boatmen arrived at the island to collect them, they found Kirwan alone and dismayed. A search of the island was made for his wife, and Maria's body was found in her bathing shift.

An inquest was opened, and as there were no signs of violence, apart from those left by crabs and the action of rocks on the sea strand, it was thought that Maria had drowned by accident. Suspicions were raised that Kirwan had drowned her: for ten years he had kept a mistress, by whom he had had seven children. What other motive could he need? There were claims that people passing near the island heard cries of distress, at the time when Maria could have died – or been killed. One woman 3 miles away from the island made the unlikely claim that she had clearly heard a woman shout 'Oh! What have I done to be murdered?' But Kirwan's sketchbook was used as evidence: how could he have captured the sunset on the other side of the island and killed his wife at the same time?

Before the case went to trial, Dr Geoghegan, one of Taylor's acquaintances, contacted him about the case. He clearly had reservations about it. Taylor thought that it was unlikely she had been drowned by her husband, especially once he found out that Maria had suffered from fits.

Kirwan's trial took place at the beginning of December, and lasted two days. In another case where question marks over the defendant's sexual morality tended to suggest guilt – as his solicitor put it, he was 'persecuted and slandered to death' – Kirwan was found guilty, accused of having drowned his wife.

On 20 December, Taylor wrote a letter in reply to John Moore Neligan, a Dublin physician, who had sent Taylor the medical evidence and asked for his opinion. In his letter, Taylor said that several of his friends doubted Kirwan's guilt; 'Mr Rodwell, the barrister, considers there was not sufficient evidence to convict' and 'my friend [Bransby] Cooper, the surgeon of Guy's, came to me and most strongly stated his opinion from the medical evidence, that the prisoner has been wrongly convicted.' Taylor said that the medical evidence could be explained by Maria having had a fit in the water; there were no marks of violence on her, which would be expected if she had been drowned by someone else, 'and that therefore to execute [Kirwan] for the alleged offence would be a most dangerous proceeding on the part of our law authorities.'

Taylor wrote an eight-page explanation of his views, considering the reports of the body's condition when it was examined the day after her death, and then after exhumation, thirty-one days later. The appearances seen a month after her death, which were attributed to murder by drowning, Taylor said were explained by decomposition in a wet grave. Taylor was having to manage with reports as he had not seen Maria's remains in person, but his arguments are clear, giving examples of other cases of drowning, asphyxiation, and strangulation. He mentioned the case of Jael Denny, and another strangling case he had worked on, when Elizabeth Pinckard had garroted her elderly mother-in-law and tried to make it look like a suicide. He included a case of homicidal drowning reported by another friend, William Edmund Image, the Suffolk surgeon who was a contemporary of Taylor's. Image often took on the analysis of suspected poisonings in Suffolk, presumably keeping up a correspondence with Taylor on the most effective methods of analysis, and like Taylor, Image was called on to examine other kinds of suspicious death. Still another friend, Dr Francis Ogston, who was a lecturer of medical jurisprudence in Aberdeen, had sent cases of drowning for Taylor to publish in the *London Medical Gazette*. Taylor extracted from these papers the detail that the condition of the subjects' hearts did not, in most cases, match that reported in the Kirwan case.

Claiming that Kirwan had murdered his wife was an assumption, as the medical evidence did not support it. 'It would be equal to declaring a man guilty of murder not upon scientific proofs, or upon data confirmed by experience, but upon mere conjecture,' Taylor said. How much Taylor's demolition of the prosecution's medical evidence played a part is not clear, but on 31 December, Kirwan's death sentence was commuted to life imprisonment.

Boswell, Kirwan's solicitor, compiled a forty-eight-page pamphlet in defence of Kirwan, with an extremely long title page. It grandly described Taylor as 'the most eminent medico-legal writer in the Empire', who claimed 'THAT "NO MURDER WAS COMMITTED"'. Even so, Kirwan served twenty-seven years on Spike Island. On his release, he joined his mistress in America.

Change of address

Taylor had written to Neligan from a new address. By the end of 1852, the Taylors had moved from the eastern edge of Regent's Park, to St James's Terrace, overlooking the park from the north, just to the west of Primrose Hill.

A sepia photograph shows a brick terrace, where each house is three or four windows wide, with a ground floor, first and second floors, and attics with dormer windows. In the 1870s, Taylor described the house as having a chandelier in one room, and that he had a north-facing library at the back of the house. Its window was 'a double ornamental sash' set 'in a bay or recess'. This is perhaps where he wrote his books and articles. They were newer houses than those on Cambridge Place, and he was much nearer London Zoo; one wonders how often he was disturbed by the trumpeting of Jumbo the elephant.

Legal poisonings?

Arsenic sales were regulated, which should have prevented unnecessary deaths. But in January 1853, two brothers from Ashford in Kent died after eating yellow cake ornaments. Twelfth Night, also known as 'Old Christmas', harking back to the position of Christmas pre-calendar reform, was celebrated with elaborate cakes. Taylor had performed the analysis on the boys' stomachs, and found arsenic: 'the yellow colour of the ornaments was produced by orpiment' (arsenic trisulphide) and the green part of the ornaments came from 'arsenite of copper, which is very poisonous even in minute doses'. At the inquest, Taylor revealed that this was not the first case of its kind that he had encountered; 'during the two last years, Dr Taylor had met with ten fatal cases from children eating these ornaments.'

He had written about Scheele's Green poisoning, which was an arsenical dye sometimes used for food colouring, in an article in *Guy's Hospital Reports* a couple of years earlier. The article had included several other sources of poisoning: red oxide of lead, hydrochloric acid, corrosive sublimate, potassium dichromate, tincture of opium, and decayed animal matter. Even apparently innocent items such as laburnum flowers and the common mussel had poisoned people. There was no regulation of poisonous food colourings: in 1848, one man had died and several more had fallen ill after eating a green blancmange at a dinner party in Northampton.

Arthur Hill Hassall, a medical practitioner, wrote a series of articles in *The Lancet* addressing the issue of food adulteration, and it wasn't just arsenic that was his focus; other adulterations included copper, lead, and iron. Despite the illness and death caused by adding these dangerous, potentially lethal items to foods, attempts for legislation were repeatedly blocked in the 1850s. The government should not interfere with freedoms, the argument went, much as it had in the lead-up to the Sale of Arsenic Regulation Act; it was up to individuals to decide what they ate. But in the late 1850s, a confectioner in Bradford accidentally used arsenic instead of plaster of Paris to pad out his sweets and killed over a dozen people. Finally, in 1860, the Act for Preventing Adulteration in Food and Drink was passed. Taylor had contributed to the pressure that would eventually make Britain a safer place.

That able analyst

The Atlee family had lived near Croydon, to the south of London. Husband, wife, and three children had died in swift succession at the end of December 1853, their deaths attributed to typhoid fever. They were a poor family, and their water came from a pond; the water was thought to have killed them. A post-mortem was performed on one of the children, and the surgeons held to their opinion that the cause of death was typhoid.

But another Croydon surgeon didn't agree; he thought it was irritant poison that had killed them. He wrote to Lord Palmerston, Home Secretary, and the mother of the Atlee family was exhumed. Taylor, 'that able analyst' as *The Lancet* described him, performed the toxicological testing. He found no arsenic in the stomach or intestines, but he found a huge quantity in the liver. Defending the medical men who had initially examined the case, Taylor said that 'to detect it in this organ requires practice and experience … and the appliances of a large and well-furnished chemical laboratory,' which was beyond the means of most surgeons.

A second inquest was held, and the conduct of the mother, and certain things she had said, suggested that she had committed suicide. There was pressure to exhume the rest of the family and have their viscera analysed; it now seemed likely that they had all been poisoned by Mrs Atlee. But the story slid out of the newspapers and medical journals, only reappearing at the end of the year to inform readers that the other exhumations never took place. Mrs Atlee had passed beyond the jurisdiction of earthly punishment.

It is possible that expense stood in the way of the Atlee deaths being further examined. James Henry Burton died in Sussex, in May 1854. Originally it was thought that he had died of natural causes, but suspicions arose and he was exhumed. Taylor performed an analysis and found a tenth of a grain of arsenic in his viscera, but the coroner declared that in order to save costs, he would read out Taylor's report at the inquest, rather than suffer Taylor's travel expenses.

Burton had been engaged to be married, but he had told a labourer whom he worked with that he'd 'got into trouble' and suggested that they ran away to sea. Had Burton's fiancée found out his plans, and killed him? The inquest jury weren't convinced and returned an open verdict; there was the possibility that poor Burton had taken his own life. There were some questions that even Alfred Swaine Taylor could not answer.

Chapter 11

Romantic, Mysterious, and Singular
1854–55

A tremendous explosion

On 6 October 1854, a fire started in a newly built worsted mill in Gateshead, a large town on the river Tyne in north-east England. The alarm was raised, and despite it being late at night, the streets were soon full of people trying to save the building. The fire had begun on the top floor, so efforts were made to rescue stock from the lower floors, but oil kept on the premises only served to feed the flames. In two hours, the roof had fallen in and the building was a wreck.

A seven-storey-high bond warehouse stood nearby, and that night it stored nearly 3,000 tons of highly combustible sulphur, 128 tons of sodium nitrate, and items such as rags, guano, soap, cement and lead; even 5 tons of arsenic. The sulphur started to melt in the intense heat and streamed like lava from the windows. The fire brigade, helped by soldiers from the nearby barracks, rushed to save the warehouse, but at three o'clock that morning, the entire structure was engulfed in flames.

The first warning that this wasn't any ordinary fire was when three bangs were heard. The firemen continued to work, and many people remained nearby to watch. Suddenly, 'the vaults of the warehouse were burst open with a tremendous and terrific explosion.' It was so enormous that boats on the Tyne were thrown about, and miners 11 miles away in the depths of Monkwearmouth, Britain's deepest pit, heard it and came up to the surface to see what had happened. Huge stones and rocks were flung about, the parish church was destroyed with the clock fixed at ten minutes past three, 'massive walls were crumbled into heaps – blocks of houses tumbled into ruins – the buildings at the confines of the towns were shaken as by the heavings of a pending earthquake.' The explosion left a massive crater, and fire raged through Gateshead. It leapt over the river to blaze its way through Newcastle, the town on the opposite bank. The dead, dying and wounded lay in the streets, as medical attendants rushed to their aid. More than fifty people were killed by the explosion, and many more were maimed.

Incredible stories of heroism emerged, not least that of Catherine O'Brien, who disguised herself in the apparel 'of the stronger sex' and worked on the fire engines during the blaze. 'She is said to have been one of the most powerful and efficient hands engaged, and no one at the time suspected her sex.'

The Great Fire of Newcastle and Gateshead was so terrible that Home Secretary Lord Palmerston intervened in the ensuing inquiry. He appointed Taylor to investigate the chemical causes of the explosion.

Taylor journeyed nearly 300 miles north. He 'visited the site of the explosion, and accurately inspected the locality; and had collected samples of the substances from various parts of the premises, including stones, timber, water, &c.' It was believed that the explosion had been caused by gunpowder, because what else could cause such an enormous blast? It was used by gunsmiths, and in the mining industry, and there were reports of shipments of it having recently arrived in Gateshead, so it was a plausible theory. But the warehouseman swore that none had been stored on his premises.

Taylor made his analysis. The substance he collected at the site 'was of a snuff-brown colour'; in 1854, everyone knew exactly what that meant. It smelt like 'recently burnt gunpowder', giving off 'a slightly offensive smell' of hydrogen sulphide gas. It contained sulphur, compounds of sulphur, and iron oxide, but it was 'assuredly not' gunpowder.

Taylor analysed the post-mortem appearances of those killed in the explosion. There was no burning caused by gunpowder on the bodies of the dead, no blackening of their faces from fine soot, 'no burning or appearances of powder' on their clothes. When the owner of a chemical works put forward the theory that the explosion was the result of steam, one of Taylor's reasons for rejecting the idea was because there were no signs of scalding on the victims' bodies.

Taylor theorised that sulphurous acid and nitrogen had been evolved as the sulphur and sodium nitrate burned. Trapped inside the warehouse vault, it had caused an intolerable build-up of pressure, which resulted in the explosion. He calculated the volume of space required to accommodate the volume of gas, which, he said, led to a pressure of 5,700lbs per square inch: it really was no wonder that the building had exploded so dramatically.

A reward of £120 was put up by the Corporation of Gateshead and by several insurance companies, to anyone who could prove that the explosion was caused by gunpowder. Vicious letters appeared in the press from people shredding Taylor's argument to pieces in an effort to prove that gunpowder was to blame. They were willing to undermine his reputation if it could earn them £120, but no one could disprove his conclusions.

The Mysterious Poisoning at Finchley

Just before Christmas 1854, 60-year-old widower John Southgate died in Finchley, London. His daughter had found four packets of Epsom salts beside his bed. The salts were used as a laxative and were only deadly when taken in vast quantities, but his surgeon believed he had died from the effects of poison. He discovered that the salts were in fact oxalic acid, used for cleaning and bleaching; it has a similar appearance to Epsom salts.

Accidental oxalic acid poisonings were not unknown, but as Taylor explained in *On Poisons*, 'its intensely acid taste, which could not be easily concealed by admixture with any common article of food, would infallibly lead to detection long before a fatal dose had been swallowed.' Death was most likely to come about by suicide; oxalic acid was rarely used by murderers. But rumour was afloat in Finchley that Southgate had been murdered, because 'he had recently made a will disinheriting some or one of the members of his family.' His life was insured for £4,500, which might also have provided someone in his family with a motive.

An inquest was opened, with Thomas Wakley, founder of *The Lancet*, as coroner. The initial theory was that Southgate had been sold the acid by accident at Neill's druggist's shop. But the assistants were adamant that they could not have muddled up the two items: the Epsom salts were kept loose in a drawer, whereas the oxalic acid was kept in a bottle on a shelf. A chemist called Mr Lucas, who had seen the newspaper reports, came forward voluntarily to say that he had sold Southgate some oxalic acid on 15 December, the day before he died. Over the past twenty years, Lucas had sold Southgate many drugs, often selling him oxalic acid because he used it in his business as a warehouseman.

Taylor examined the oxalic acid found in Southgate's house, and a sample from Lucas because, even on 'opening the packets, the two specimens did not at all agree in appearance.' Taylor was able to show that, chemically, they were not the same, and the oxalic acid from Neill's shop was different again. Taylor's analysis went further: using his microscope, he examined the fresh parcel of oxalic acid that had been cut open by Wakley at one of the inquest meetings. Taylor was of the opinion that it had not been tampered with. Proving that no one had maliciously interfered with it could rule out murder.

Letters appeared, which Southgate was secretly writing to a mysterious 'J.M.'. His tenant, who took the replies in for him, revealed that J.M. was a young woman. The police discovered that she was 18-year-old Jane Marden, Southgate's former servant; one of the reporters described her as 'an intelligent and interesting-looking young woman'. They had been seen together at a hotel. Southgate professed that he would marry her, but could not while his two daughters were unmarried.

Southgate had been a wealthy man, but after the police investigated his business deals, his warehouseman presented the inquest with an accounts book full of large sums that had been paid out to people only recorded by their initials. £90 had gone to one J. Marden; Jane denied that it was her, and said that the signature wasn't in her handwriting. It was not mentioned in any of the reports, but Jane had a brother called James: was he blackmailing Southgate?

A merchant came forward to say that he had lent money to Southgate, and that he often sold Southgate perfumery items. On 2 December, Southgate had bought prussic acid and essential oil of almonds from him. Southgate's will was dated 4 December, two days after he had bought the two potentially poisonous substances; it seems as if he was planning his own demise in the manner of a careful businessman.

The inquest, with its chemical and romantic twists and turns, lasted for six meetings. It had fascinated the press, who described it as a 'romantic, mysterious [and] singular inquiry'. Even the Home Secretary had taken an interest. The jury spent two hours deliberating and finally decided that Southgate had 'died from the effects of oxalic acid, administered by his own hand while in an unsound state of mind'. But where that poison had come from, no one ever found out.

Poison is there

After the Southgate investigation, Taylor was asked to analyse the water in Cherry Hinton, Cambridgeshire, by the appropriately named Mr Peed. Taylor had begun public water analysis while working with Aikin, as it required their chemical expertise. Towns were proud to announce that no other man than the eminent, celebrated Professor Taylor was to analyse their water supply. He could identify if there was unsafe 'organic matter' floating about in it, as well as pollutants that were the run-off from industry. That towns were taking water safety seriously was a good sign, and would have gladdened Taylor, for whom public health was important. But while a water supply could endanger the lives of a whole town, the public were gripped by crimes that homed in on one household.

In August 1855, another middle-class poisoning came to light. Joseph Snaith Wooler lived with his wife, Jane, in Great Burdon in north-east England, about 40 miles south of Newcastle. The Woolers had been married for nearly twenty years, and were just as affectionate as they always had been. So much in love were they that they had, somewhat unusually, married twice: first in 1836 at St Martin's-in-the-Field's in London, Wooler's home city, and again in 1837 in Wolsingham, on Jane's home turf. For a time, they had lived in India, where Wooler had been a merchant, but they had returned to England, and now Wooler's considerable income came from property.

Although Jane's health had never been good, she had not been seriously ill until early May 1855, when she suffered from sickness and diarrhoea. She was cared for at home by two ladies of her acquaintance, as well as by Ann Taylor, the Woolers' servant, and three doctors who had trained at Edinburgh Medical School – Dr Jackson, his assistant Dr Henzell, and Dr Haslewood. The doctors initially thought that Jane was suffering from tuberculosis. Her husband, who evinced a 'very intimate knowledge of medicines', helped to administer medicines to his wife, including 'injections' by enema syringe, aided by Ann and the attending ladies.

The doctors had been taught by Robert Christison, and it might have been his influence that led to the surgeons growing suspicious. In early June, Jane began to suffer extra symptoms, such as intense thirst and tingling in her hands, which made them suspect that she was being poisoned with arsenic. Jackson and Henzell eventually discussed their fears, and when they recalled that her husband knew a lot about medicines and had a large basket full of them, amongst which was a bottle of the arsenical tonic Fowler's Solution, it seemed they had a culprit.

'Poison is there,' Jackson said, 'and there is some person in the house who understands the use of it.'

But Wooler was a wealthy man, a fine, upstanding gentleman. He was no Mary Anne Geering; he couldn't be dragged out of a cottage by a police constable. They had to proceed with care.

They wrote to Christison for advice, and their former tutor replied:

Edinburgh 25 June
Dear Sir,
I received this morning the detailed statement of your perplexing case, and I hasten to give you what information I can.

It appears to me either that the moral evidence must be very strong, or that arsenic must be detected in remains of food, vomited matter, or in the excretions, before it would be prudent to avow your suspicions of slow arsenical poisoning. I cannot well judge of the force of the points of moral evidence you refer to in your letter, but you will not, I hope, think it wrong in me to remind you that when an individual comes under suspicion in such peculiar circumstances, many natural acts are apt to put on an unfavourable appearance to the prepossessed observer.

The doctors had already started to administer what they hoped would be an antidote. They noticed, too, that on the day that Wooler was away from home, Jane's condition improved; on his return, it worsened. Henzell had been analysing her urine daily for indications of disease, and one day noticed that the urine had been substituted for someone else's. It was only in late June that he thought to examine the daily urine samples for arsenic; they tested positive, as had the sample they had sent to Christison. On 27 June, Jane Wooler died of tetanic spasms, which can sometimes result from long-term arsenic poisoning.

Jackson refused to give a death certificate without an inquest. There was a post-mortem, with Wooler's consent, but a portion of the intestines was removed for analysis, without the knowledge of Jane's family. The physician who performed the analysis found arsenic. At the inquest, Jackson voiced his suspicions about Wooler and his basket of medicines, and Wooler complained that Jackson had kept his suspicions to himself; 'He had ample means, and would have given a thousand pounds to save his wife.' The coroner thought there was 'mystery in the case', but not against Wooler or anyone in particular. The jury returned an open verdict: Jane Wooler had died of irritant poisoning, but there was no evidence to say by whom the poison had been administered.

Jane's family were not satisfied, and Wooler was arrested. An investigation was opened before the county magistrates, which ran through most of August. Wooler co-operated; he wanted an investigation, 'for the sake of his own character and the character of his family'. His legal representatives pushed for an exhumation, as they were 'dissatisfied with the manner in which the first portion [of viscera] had been removed'. They wanted a second opinion, and would pay 'Professor Taylor or Mr Herapath, or some other eminent man, for analysis'. On 10 August, after the viscera had made a journey of about 250 miles, Taylor began his meticulous analysis with his colleague, George Owen Rees.

Taylor examined the liver with a magnifying glass, and on finding a tiny piece of newspaper adhering to it, he removed it and tested the newspaper for arsenic. He wanted to be sure that if he did find any arsenic in the viscera, it had not got there by blunder. He ensured that the hydrochloric acid for the Reinsch test was pure by making his own; this was something he had done for years, to avoid the risk of it being contaminated with arsenic. They were thorough and careful. Taylor and Rees examined half of the heart each, and then swapped, 'to check each other'.

They found arsenic in every part of the body that they had been sent: the heart, liver, intestines, and part of the lungs, but had not found any in the blood from the abdomen. Yet the arsenic was in small quantities. In the liver, heart and lungs, 'the arsenic was incorporated with the structure of those organs' and so must have been administered during life. Taylor found that Mrs Wooler had been suffering from liver disease, the appearance that of 'a person living in a tropical climate, or in persons labouring under consumption'. There was inflammation and ulceration in the intestines, but it could have been caused by arsenic or disease. Taylor was fairly sure that the arsenic hadn't been taken as a powder, but in liquid form, 'whether in water, milk, or in the form of Fowler's Solution', he couldn't say.

Only one grain of arsenic had been found in total. It was not enough to kill when taken on one occasion, but it was what remained after having been administered in small amounts over a long period of time.

At the end of August, Taylor made the journey north and explained his analysis at the inquest.

Dr Fothergill, another local doctor who had been drawn into the case, gave evidence about the enema syringe he had lent Wooler. Fothergill tested it, wondering if it could have been used to administer the arsenic. He did find a quantity of it present, but then realised that the hydrochloric acid he had used was impure, although not enough to explain the quantity of the arsenic he had found in the syringe.

Taylor had known about this in advance; Henzell had asked if they should tell the magistrates. Taylor replied that they 'should without delay'. He disliked any uncertainty, and a delay could cast expert witnesses in a bad light.

After Fothergill had explained the issue of his impure acid to the inquest, Taylor detoured into a speech on the adulteration of food. At the time, it was thought by some that unfermented bread was healthier than bread made with yeast; unfermented bread used flour, soda, water, and weak hydrochloric acid. Considering the risk that the acid could contain arsenic, Taylor declared that 'it was disgraceful that so dangerous a substance was allowed to be sold'. He had been about to communicate this issue to the House of Commons committee inquiring into food adulteration, when he had been called to the inquest. By mentioning this shocking fact at an inquest that was receiving column inches in newspapers up and down the country, Taylor was ensuring that the public would get to hear about this 'exceedingly dangerous and pernicious' practice.

This was the last meeting of a long, drawn-out investigation; it was decided that Wooler would stand trial.

Excited by drink

In the months that intervened, Taylor's opinion was sought in a messy paternity suit. They were painful cases, where a child's legitimacy and its mother's morality were tested by the law. Ann Matthews had married Richard Legge in 1837, in the Gloucestershire village of Newent. He was a drunk, and one child was born in 1839, which did not survive. Richard 'suffered severely from early excesses' and in November 1843 he was struck with hemiplegia – paralysis affecting one side of the body. He was confined to his bed, but was eventually able to get about in a wheelchair, and then on a crutch. Conflicting reports said that he had been seen on horseback.

During Richard's illness, a solicitor called Edmund Edmonds was often at the house, sorting out the intemperate man's muddled finances. Rumour had got about that Edmonds was closer than he should have been to Richard's wife. At the end of January 1844, Richard was seen apparently recovered at a Methodist tea party where, 'excited by drink', he challenged someone to a fight, but days later he was attacked by paralysis again, and he died in June. That October, Ann gave birth to a daughter, Mary Frances, and Edmund Edmonds married Richard's widow.

Richard's mother died intestate in 1845; he had two surviving siblings, so his daughter inherited what would have been Richard's third portion of his mother's estate. The child died in 1849, so the inheritance passed to Ann and Edmund. Richard's siblings challenged this in Chancery, claiming that Richard was impotent; he couldn't possibly have fathered a child, so Mary Frances's inheritance was theirs.

There were witnesses – or at least, gossips – who would attest to Ann and Edmund having committed adultery, but there were also letters that Ann had written before her marriage to Edmund, which implied that the child was actually his. Richard and Ann had been married for eight years, with only one – or two – children being born to them, but following Ann's marriage to Edmund, she had given birth to several children in quick succession. It seemed that Mary Frances could well have been Edmund's daughter.

Legally, a child is its father's if the parents are married and have access to each other, but there were legal precedents, established to protect property from the offspring of philandering wives. Could medical science help? Taylor was one of four doctors who worked on the case. Ordinarily, they would look for physical similarity between father and child, but neither Richard nor Mary Frances were alive, so this was impossible to establish. Taylor and Dr Carpenter reasoned that it was possible for Mary Frances to have been Richard's daughter, but not probable. However, two other physicians, Dr Semple, and Dr Guy – a man against whom Taylor may still have born a grudge – said that it was quite possible for Richard to have fathered the child.

The Hereford Times reported the case, which spilled across two weeks' worth of issues, but perhaps it served to put rumour to rest. The arguments of the barristers were printed in depth, but the medical evidence was suppressed from public view, it being deemed rather racy as it discussed in detail whether or not a man with hemiplegia could perform sexually. The judge said that if they couldn't prove that Richard was impotent, then they would have to accept that Mary Frances was his child. The compromising letters were overcome by declaring that, on various grounds, they were inadmissible. The case was found in the defendants' favour. Taylor avoided mentioning his failed case in his medical jurisprudence books, but Guy happily included it in his own.

My tests were pure

Wooler's trial at Durham that December lasted three days. Christison and Taylor both gave evidence; it is likely that this was the first time the two men had met.

The medical evidence was of extreme importance, with Mrs Wooler's doctors being examined, as well as Britain's two leading lights of medical jurisprudence, Taylor and Christison. All of the medical experts agreed that Mrs Wooler had died of arsenic poisoning, even though her symptoms had not initially suggested it.

Serjeant-at-Law Charles Wilkins led Wooler's defence. He was considered to be a valuable barrister in poison trials as he was a surgeon's son who had trained in medicine. But he had run away from his apothecary apprenticeship, turning to law in his late twenties after a career on the stage. Although he had some medical knowledge, could it match that of practising medical professionals?

Between the inquest and the trial, Taylor had been sent seventy bottles of medicine from Wooler's house, and three enema syringes. He didn't find arsenic in any of the bottles – whatever happened to the Fowler's Solution would forever remain a mystery – but he did find traces in one of the syringes. Even so, he said there was no evidence to say that arsenic had been administered by the syringe; he believed that it would be a very ineffective way to administer it, as the body would reject it quickly before it had time to be absorbed. In answer to a question about possible contamination of hydrochloric acid, Taylor replied, 'My tests were pure.'

Wilkins's closing speech blamed Mrs Wooler's doctors for her death, using some of Taylor's words to do so, even though Taylor had said that the initial symptoms they had been faced with did not point exclusively to arsenic poisoning. Wilkins declared, Taylor 'says that the poison might have been administered "or taken"', which Wilkins took to suggest that Mrs Wooler had been given impure or adulterated medicine. He accused the doctors of negligence for not acting on their suspicions at once. The judge, Baron Martin, summed up. He didn't take as strong a view as Wilkins, but still declared the doctors' conduct to have been 'reprehensible', criticising them for not taking action earlier. He ended with the enigmatic statement, 'If I were to make a surmise, there is a person upon whom my fancy would rest rather than upon the prisoner.'

The jury acquitted Wooler. Baron Martin said he would have interfered sooner, presumably to throw the case out, but he 'thought it more satisfactory to allow the case to be fully heard'. Thus all of the evidence was brought out and Wooler was properly cleared before the public, removing any lingering suspicions.

And yet the mysteries surrounding the case didn't go away. Wilkins had wanted to claim in his defence that Mrs Wooler was an arsenic eater, based on a letter written the year before by Swiss naturalist Von Tschudi, in a journal called *The Chemist*. He claimed that Styrian peasants took regular doses of arsenic to improve their complexions and help them live at altitude, and that they started with small doses of half a grain and gradually increased it. Wilkins had wanted to claim that, like an opium eater, Mrs Wooler had died because she had been denied access to the arsenic that she had become habituated to taking. Taylor believed that the medical evidence did not show this at all, as there was still arsenic in her system, being flushed out in her urine just before her death. An anonymous letter was written to *The Times* to suggest the arsenic-eater theory, but the *Pharmaceutical Journal* thought it as 'equally absurd' as the idea that the arsenic found in Mrs Wooler's body was naturally occurring. They were dangerous theories that might allow the guilty to get away with murder.

Privately, a judge, James Parke, contacted Taylor with two questions connected to the case. In a personal letter to another judge, Parke commented on Taylor's response. 'Admirable answers they are – clear, distinct, convincing. They leave no doubt in my mind that justice has not been done.' Parke admired and trusted Taylor, having shared court rooms with him before. He could not understand why Baron Martin thought that there was no case to pass to the jury, but he urged his friend's confidence in the matter: 'Don't give my opinion to him.'

Following Baron Martin's words, the public were looking for a suspect. Haslewood and Jackson, feeling accused, wrote to *The Times*. Haslewood, calling Christison 'the father of modern forensic medicine', forwarded the letter that the great man had sent, urging the doctors' caution. It hadn't been fair for them to be criticised for not taking early action, when Christison had advised them not to do so. Jackson sent in a letter he had written to Baron Martin, including the judge's reply. 'Your Lordship's words may mean either that I gave the deceased poison wilfully or through culpable negligence.' Baron Martin said that he had meant, in strictly theoretical terms, that the evidence was so weak that any 'fancy or surmise' against Wooler could equally 'be directed against another'.

The Darlington Times called it a *cause célèbre*, 'one of the most remarkable that ever occurred in the country'. Thanks, perhaps to Baron Martin's closing words, 'it is wrapped in as much mystery now, after the trial, as it has been from the first.' But the Great Burdon Poisoning was unseated almost immediately; all eyes were focused on the Midlands market town of Rugeley and a surgeon called William Palmer.

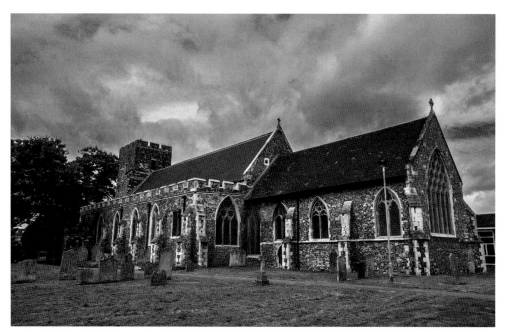

Plate 1. St Botolph's church, Northfleet, where Alfred Swaine Taylor's parents were married, and where he and his brother were baptised. Several members of Taylor's family are buried in the churchyard. (Gordon Wallace)

Plate 2. Albemarle House, Hounslow, 1804. Taylor was a pupil from 1816–22. (© British Library Board Ktop XXX, 3-a)

Plate 3. Guy's Hospital, London. (Wellcome Library, London)

Plate 4. Robert Christison. (*Illustrated Edition*)

Images from *Illustrated and Unabridged Edition of The Times Report of the Trial of William Palmer for Poisoning John Parsons Cook at Rugeley*, published by Ward & Lock in 1856, originally appeared in the *Illustrated Times*. They are indicated with *Illustrated Edition* following in brackets.

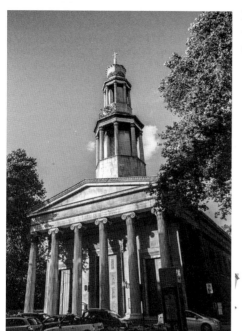

Plate 5. St Pancras church, London, where Taylor was married, and where his children were baptised. (Gordon Wallace)

Plate 6. Chester Gate, previously Cambridge Place, on Regent's Park, London. Taylor lived in a house here for nearly twenty years from 1834. (Gordon Wallace)

Plate 7. St James's, Piccadilly, where Taylor's only son was buried in 1836. (Gordon Wallace)

Plate 8. A cyanotype of lace and a feather, using the same principle as Taylor's 'photogenic drawings', but different chemicals. (Helen Barrell, with thanks to Rachel Brewster of Little Vintage Photography)

Plate 9. Granby Place, Northfleet, home to Taylor's father until 1840. (Gordon Wallace)

Plate 10. Regent's Square, London, where Taylor's brother lived in 1841. An old medicine advert is reappearing on the wall. (Gordon Wallace)

Plate 11. Portrait of Taylor, c. 1840. (Image supplied courtesy of Guy's and St Thomas' Charity)

Plate 12. Dr George Owen Rees, a colleague of Taylor's at Guy's Hospital. (Wellcome Library, London)

Plate 13. Henry Letheby.
(*Illustrated Edition*)

Horrible and Bar-bar'-ous Murder of Poor
JAEL DENNY,
THE ILL-FATED VICTIM OF THOMAS DRORY.

Plate 14. A depiction of Jael Denny's murder could be bought from street sellers. It included detail from reports of her post-mortem. From Henry Mayhew's *London Labour and the London Poor*. (Cadbury Research Library: Special Collections, University of Birmingham)

Plate 15. William Palmer. The image had originally appeared in the *Illustrated Times*. (*Illustrated Life and Career of William Palmer of Rugeley*, Ward & Lock, London, 1856).

Plate 16. William Palmer's birthplace, Rugeley. (Gordon Wallace)

Plate 17. Rugeley, 1856. The Talbot is on the left, Palmer's house is on the right, behind the railings. (*Illustrated Edition*)

Plate 18. The same view of Rugeley in 2016. The front of what was once Palmer's house has been extended, and is now occupied by a pet shop and a weight-loss salon. (Gordon Wallace)

Plate 19. Palmer's trial at the Old Bailey. (*Illustrated Edition*)

Plate 20. William Herapath. (*Illustrated Edition*)

Plate 21. Taylor and Rees in the laboratory – or were they? (*Illustrated Edition*)

Plate 22. William Ward Webb, the compromised coroner. (*Illustrated Edition*)

Plate 23. George Bate. (*Illustrated Edition*)

Plate 23. George Bate. (*Illustrated Edition*)

Plate 24. Originally appearing in the *Illustrated Times* to depict the inquest into Walter Palmer's death, this image was repurposed by the *Illustrated and Unabridged Edition* … as the inquest into John Parsons Cook's death. (*Illustrated Edition*)

Plate 25. Thomas Nunneley. (*Illustrated Edition*)

Plate 26. Elizabeth Mills gives evidence at Palmer's trial. (*Illustrated Edition*)

Plate 27. Cook's grave, Rugeley, 1856. (*Illustrated Edition*)

Plate 28. Cook's grave, Rugeley, 2016. (Gordon Wallace)

MURDER WILL OUT

"Foul deeds will rise
Tho' all the earth
O'erwhelm them from man's eyes."

ALL CLASSES SHOULD READ THE

Midnight Deeds & Mysterious Doings
at MADAME DEVEREUX'S
DEN OF INFAMY,
IN GRANBY ST., WATERLOO ROAD,

Where the Vilest Assassins of London assemble to Plunder & Murder the unfortunate victims who are entrapped within

those walls of Infamy by the frail Sisters, the Infesters of the night who frequent the Waterloo Road and its neighbouring Streets.

IN THE ABOVE HOUSE ARE

GAMBLING ROOMS, THE SULTAN'S HAREM, THE LEWD ROOM, THE IRON ROOM,
WHICH CONTAINS A HORRIBLE INSTRUMENT OF TORTURE CALLED

THE MAN-COG,
AND A VAULT OF DEATH,

Showing in a plain, unmistakable manner, the probable Fate and possible Clue to the perpetrators of the

Waterloo Bridge Tragedy.

INCLUDING

Ideas of Prostitution; its Causes and its Remedies.
BY THE REV. C. H. SPURGEON,
MINISTER OF PARK STREET CHAPEL.

Printed and Published by W. DE FR J8 Great St. Andrew Street, Broad Street. Bloomsbury.

Plate 29. Reverend Spurgeon's broadside alleged that the victim of the 1857 Waterloo Bridge Mystery had been murdered in a nearby brothel. (© British Library Board 1888.c.3(63))

Plate 30. Taylor and his friend, chemist William Thomas Brande, c.1863, when their textbook *Chemistry* was published. (Wellcome Library, London)

Plate 31. Taylor poses for Ernest Edwards in his studio on Baker Street, London, 1868. (Wellcome Library, London)

Plate 32. Taylor photographed by Barraud and Jerrard, 1873. (Wellcome Library, London)

Chapter 12

Enter Not into the Path of the Wicked
1855–57

Mysterious death of a sporting gentleman

At the end of November 1855, chemist Robert Warrington introduced Taylor to the recently bereaved William Vernon Stephens. It was just over a week before the Wooler trial, but Warrington was a 'chemist of well-known and considerable ability' himself, and presumably recommended Taylor as the best person from whom Stephens could seek advice. Accompanying Stephens and Warrington was the clerk of Rugeley solicitor Mr Gardner. They brought with them a jar containing the viscera of Stephens's recently deceased stepson, John Parsons Cook.

Cook had been only very young when his merchant father died intestate in 1831. There immediately followed a Chancery dispute between Cook and his own mother, instigated by Cook's uncle. In 1838, Cook's mother married Stephens, a gentleman shareholder, but in May the next year she died, apparently in childbirth. When Cook's grandfather died in 1845, he left Cook, his sister and his half-brother the rents and income on various properties when they came of age. About this time, Cook was articled to a solicitor in Sussex, but rather than focus on his law career, he enjoyed theatre trips, dinner parties, billiards, 'the less dangerous fascination of cricket', races and regattas.

He lived near Goodwood race track and famous training stables. By the time he left Sussex in 1850, Cook was on his way to becoming a betting man. In 1851, he appears to have used his income from his grandfather to go into farming, with over 100 acres and a team of labourers toiling the soil for him. He lived in style for a single man in his early twenties: he had a housekeeper, a cook, a housemaid, and a footman, although his shepherd lived in with him. He could have led a respectable bourgeois existence, but the allure of drink, fast women, and faster horses was too much for him.

By 1855, Cook kept racehorses. His finances had taken a bad turn and he was a lodger in the house of a surgeon called William Henry Jones. On 13 November, his luck changed when his horse, Polestar, won a race at the Shrewsbury Autumn Races. Cook had a large circle of friends and was kind-hearted: to celebrate Polestar's win and the large amount of money he suddenly found in his pocketbook, 'he gave a dinner at the Raven, at Shrewsbury, and treated his guests to foaming beakers of provincial champagne.'

But there was at least one among that number who had reason to be jealous. Just as Cook had abandoned his legal career for a life on the turf, William Palmer was a young man who had given up his career as a surgeon to follow the horses; he left it to an assistant to run his surgery. Palmer's stable of seventeen racehorses required a huge amount of money to maintain, and he was a heavy gambler. To add to his woes, he was being blackmailed for some compromising letters. His debts were vast, about £25,000, woven into a confusing web between different creditors, with some loans secretly taken out in the name of his elderly mother. It was not uncommon for racing men to share ownership of horses, as well as betting for each other, which increased Palmer's financial complications; who, for instance, had the authority to collect the winnings? That November, Palmer was in desperate financial straits, his creditors demanding money that he simply didn't have. But if he could get his hands on some cash, it would cover his most pressing debts and give him some breathing space.

As Cook and Palmer sat together drinking, a racing friend called Ishmael Fisher heard Cook remark of his brandy, 'There is something in it – it burns my throat awfully.' Palmer refused to believe him. There was a tiny amount left in the bottom of the glass, so Palmer drank it, and then asked Fisher if he thought there was anything in it. Later, Cook told Fisher that he 'had been to the watercloset, and had been dreadfully sick, and that he believed Mr Palmer had "dosed" him.' Cook gave Fisher about £700 or £800 for safekeeping, convinced that he had been '"dosed" for the sake of the money'. The next morning, Cook challenged Palmer, who denied all knowledge of it, and Cook accepted his word.

Palmer's horse The Chicken ran two days later. It failed to match Polestar's dazzling performance, coming in fourth. That day, Palmer went home to Rugeley, Cook accompanying him. Cook had visited Rugeley before; he stayed at the Talbot Arms, the pub opposite Palmer's house. Just after midnight, on Wednesday, 21 November, 28-year-old John Parsons Cook died.

A horror of bad smells

How much Taylor had been told of Cook's background by his stepfather is unclear. He had been informed that Cook had been 'of average health', which skipped around the rumour that Cook had suffered from syphilis. Certainly Cook's death had been sudden. Taylor was given a brief sketch of Cook's symptoms; of his attack of 'nausea and sickness, but without purging' in Shrewsbury, and that he had suffered 'a fit' on the Monday night in Rugeley. On the Tuesday, 'about midnight, he was seized with convulsions, and went off suddenly.' It was thought that 'he had taken a pill, or some medicine which had been prescribed for him,' and Taylor was told that 'an aged medical practitioner had been in attendance upon him shortly before his death.' Taylor may have considered that, if death wasn't due to natural causes, then it was due to the elderly surgeon making a dispensing error.

Taylor probably knew nothing of William Palmer at the time. As the case involved a medical professional, Taylor 'requested that my colleague, Dr Rees, should be associated with me in the analysis, so that, for the satisfaction of all parties, there might be a witness to the proceedings.' Stephens agreed, although such a procedure was certainly not unusual, as Rees had worked with Taylor on the Wooler case. But when writing about the case for posterity, Taylor wanted everyone to know that he had acted in a manner that was entirely above reproof.

Rees had been born in Smyrna, Turkey, the son of a British consul and merchant, and his Italian wife. He gained his MD from Glasgow Medical School, and when his strengths as a chemist were identified, he worked at Guy's with Dr Bright on kidneys and blood. He went on to hold important medical posts, and was friends with Roget of the thesaurus. He was small and slight, so when he and the unusually tall Taylor walked along the corridors of Guy's together, they must have appeared a remarkable pair. Rees was a dandy with highly polished boots, who poured 'awful imprecations' on a patient who dared to touch his hat. He avoided the outpatients, and had 'a horror of bad smells', doing what he could to avoid post-mortems. If he was forced to attend one, 'his visit would not extend beyond the door, where he would stand for a few moments with ears open but with fingers to nose.'

On 28 November, Taylor and Rees began their chemical examination of Cook's viscera. 'There was a feculent odour observed,' Taylor adroitly put it, which Rees no doubt recoiled at. The jar contained only the stomach and intestines. The stomach had been cut open, sliced from one end to the other, and turned inside out; rather than being separated from the intestines, as was usual practice, it was lying on top of them. 'The stomach contents (if any) had entirely drained away from it.' Taylor couldn't tell if the stomach contents were somewhere in the jar, mixed up in the mess of the intestines, or if they'd even made it into the jar in the first place. It was difficult to proceed with what they had. Taylor applied for the remainder of the viscera, and the liver, kidneys, and spleen arrived with a small bottle of blood on 1 December.

The feculent odour masked any smell of opium, prussic acid or 'ardent spirits', had they been present. In the absence of stomach contents, they made do with the washings of the stomach; poisons like arsenic would get caught up in the mucus lining. They tested for the metallic poisons, using the Reinsch test, and found a small amount of antimony. Then they tested the stomach's coats again for inorganic poisons, but also for organic ones: prussic acid, oxalic acid, opium, morphia, strychnia, veratria, nicotina, and hemlock. Once again, the only thing they found was a tiny quantity of antimony, and the same was true of the other viscera, and the blood. They drained out the jar that the stomach and intestines had been in, hoping that the liquid would contain the stomach contents, and here they found the largest proportion of antimony.

Taylor wrote to Gardner, requesting him to ask Cook's medical attendant if he had prescribed antimony. Gardner replied on 2 December to say that he hadn't. It was in this reply, or perhaps earlier, that Taylor would find out that it wasn't the elderly doctor at all who had slipped up; without naming William Palmer, Gardner told Taylor that 'the party who is suspected' had been in the medical profession, but had given it up. He informed Taylor that the day before Cook's death, the suspect had bought strychnine, prussic acid, and liquid opium.

Taylor and Rees wrote up their conclusions on 4 December: they saw nothing in the condition of the viscera to have caused a natural death, and they stated that antimony, such as tartar emetic, 'may be given as a safe and innocent medicine, or be used as a poison'. Taylor told Gardner, 'Dr Rees and I have completed the analysis today,' and explained that there would be a delay due to the Wooler trial. He asked Gardner for a thorough list of all the medicines that Cook had been given. Referring to the possible poisons that Palmer was known to have bought, Taylor wrote, 'We do not find strychnine, prussic acid, or any trace of opium.' But he did discuss the problems of the stomach contents having drained away: 'It is now impossible to say whether any strychnine had or had not been given just before death.' Taylor wrote that with nothing else to suggest a cause of death, it might have been the antimony that had killed Cook.

Oh, doctor, I shall die

William Palmer pleaded ill health and did not attend the inquest. The first meeting was mainly taken up with Jones describing Cook's illness. Cook had been his lodger, and he had come to Rugeley at Palmer's request. Jones reported that Palmer, with Bamford – the elderly medical gentleman – had decided on treatment.

On 13 December, Palmer wrote a letter to William Webb Ward [Plates 22, 24], the coroner, who happened to be a friend of his. Palmer plied him with game and oysters, and sent him a £10 note. Amongst other comments on the case, Palmer wrote, 'whatever Professor Taylor may say tomorrow; he wrote from London last Tuesday week to say "We (Dr Rees and I) have this day finished our analysis and find no traces of either strychnine, prussic acid, or opium."' Palmer avoided mentioning the antimony, and ended his letter, 'I hope the verdict tomorrow will be that he died from natural causes and thus end it.'

Rugeley's postmaster later claimed that he had found Taylor's letter open in the bag of post from London, and that he was so elated to see that they hadn't found any poison, that he had rushed to tell Palmer the good news. But Palmer's letter nearly quotes Taylor word-for-word, as if he had seen the letter, or a copy of it. The postmaster was telling a fib.

The first witness to be examined when the inquest was resumed on 14 December was Bamford, who explained Cook's symptoms, but went into little detail about the medicine he had prescribed. Taylor had arrived, and asked if he could put some questions to Bamford. The coroner allowed him to, and Taylor asked Bamford what the medicine he had given Cook contained. Morphia and mercury, amongst others drugs, Bamford replied, but not, it turned out, any antimony – knowingly, at least. Bamford was convinced that Cook had died of congestion of the brain, and he believed that the post-mortem proved it. But his examination, rather than deflect suspicion from Palmer, tended to increase it, when he answered a question about attending other patients for Palmer. There had been Palmer's wife, and two children, and a man who had visited from London: all of them had died.

Taylor explained the tests he had carried out with Rees, and the conclusions they had reached; he avoided any comment on the condition in which he had received the viscera. When questioned about the antimony, Taylor said that it must have been taken 'at least eighteen hours before death', perhaps even days earlier. Between eighteen to twenty grains was a lethal dose, so medicinal quantities were not fatal, unless it caused sickness, leading to exhaustion. It could cause convulsions in large doses. Before giving an opinion on the cause of death, Taylor asked to hear Jones's evidence, which he had missed; after all, Jones was a medical man and he had been present when Cook died. The coroner read it over, 'Dr Taylor taking notes of it as he proceeded, and frequently interrogating Mr Jones with reference to the precise time that certain symptoms were observed.'

The coroner asked Taylor what would cause Cook's heart to be emptied of blood, and Taylor said this would have been down to the spasmodic action of the body at death, 'caused by poison or disease'.

Taylor listened to the eyewitness account of Elizabeth Mills, a chambermaid at the Talbot Arms. [Plate 26] This, combined with Jones's evidence, gave Taylor an idea of Cook's last days, although it wasn't until several weeks after the inquest that he was able to piece it together in full.

Cook had arrived in Rugeley unwell, and went to bed at 10.30 pm. He dined at Palmer's on the Friday, and went to bed that night sober, but on Saturday and Sunday he could not get up at all. Palmer sent Cook some broth on Saturday, but Cook wouldn't eat it, until Palmer took it up to him. Elizabeth later found it 'in the chamber utensil, as if it had been thrown off the stomach'.

He was still weak on Monday, but the sickness ceased, and he managed to eat. Palmer sat up with him that evening, but had gone home by the time Cook called Elizabeth, at about midnight. She found Cook in his bed, beating the bed with his hands. Cook said his illness had been caused by some pills he had been given by Palmer at 10.30 pm. His arms and legs suddenly 'became perfectly straight' and Cook said to Palmer, who had returned to the pub, 'Oh, doctor, I shall die.'

Palmer replied, 'Oh, no, my lad, you won't.'

Palmer brought two pills, and 'some mixture of a dark thick kind, in a wine glass; it smelt like opium'. Cook took the pills, then the contents of the glass. He vomited, but they couldn't find the pills. Elizabeth sat up with Cook until three o'clock in the morning, and he asked her to rub his stiffened hands.

Jones's evidence was from the Tuesday night. Palmer had written to him to say that Cook was ill with 'a very severe bilious attack', and recommended he attend. Jones examined Cook and disagreed with Palmer's diagnosis, thinking his tongue was too clean. 'You should have seen his tongue before,' Palmer replied. Jones checked on Cook every half an hour. Bamford called at about 7.00 pm, and he and Palmer decided that Cook should be given more morphine pills, but that he shouldn't be told, because they had made him ill the night before. At 11.00 pm, Palmer arrived with more pills, and 'Cook made strong protestations against taking them, saying that he was certain they made him ill the night before.' He threw up almost as soon as he had taken them, but Jones and Palmer couldn't find the pills, so presumably Cook had managed to keep them down.

Cook's symptoms began an hour later.

'Doctor, get up; I am going to be ill; ring the bell for Mr Palmer,' he said.

Jones, who was sleeping in the same room as Cook to keep an eye on him, sent for Palmer, who arrived in two minutes. Palmer commented that he had never dressed so quickly in his life. Jones wondered if he had undressed for bed at all. He had two pills with him, which he said were ammonia – peculiar, Jones thought, as he had never known a medical man to keep such pills ready made up. As soon as Cook took them, 'he uttered loud screams, and threw himself back on the bed in very strong convulsions.' He asked to be lifted up and said, 'I shall be suffocated.'

Jones and Palmer couldn't lift him, so they turned Cook onto his side. Jones listened to Cook's heart, 'which I found to gradually cease.' In a few minutes, his friend died. Jones had never known Cook to have a fit before, and he had never seen such strong symptoms, either. He disagreed with the elderly Bamford: 'They were symptoms of convulsions and tetanus, every muscle of the body was stiffened.' The jaw was 'fixed and closed', and his body 'was stretched out, and resting on his head and heels'.

Taylor had one more question. Did Cook have any external wounds or lacerations that could have caused tetanus? The reply was no.

Taylor was ready to give his opinion. 'My belief is that he died from tetanus, and that that tetanus was caused by medicine given him shortly before his death.'

He meant tetanus in physiological terms; the tetanic convulsions that were the symptoms of infection or poison. As there were no wounds, infection was ruled out: it had to be poison.

Taylor explained his reasoning. The medicine described by Bamford couldn't have caused the symptoms, and Taylor had found no morphia or mercury in Cook's remains. This meant that on the Monday and Tuesday nights, Bamford's pills had not been given to Cook. Mills had described the symptoms of a small dose of strychnine, and Jones had described those of a large dose. Tetanic convulsions could occur in other poisonings; Jane Wooler had died of them, but this was after a long exposure to small doses of arsenic, and her other symptoms had pointed to arsenic. Taylor didn't think it could be prussic acid, the effects of which almost always came on at once; the longest they might be delayed for would be half an hour.

The problem was, Taylor explained, the chemical analysis. Arsenic 'and poisons of that kind' – inorganic poisons – 'would remain in the body and bear the test of chymical analysis.' But strychnine was different. It 'was so speedily absorbed in the blood that in the course of an hour after administration no chymical test at present known could detect it.' He didn't add that the viscera were in such poor condition when he received them, that it was a wonder he could find anything.

In the most recent edition of *Manual of Medical Jurisprudence*, Taylor had mentioned strychnine, but had focused on nux vomica, which contains strychnine, and which he had personal experience of in the Wren case. He would have remembered the vomiting immediately brought on by nux vomica's bitterness, echoed by Cook vomiting each time Palmer gave him his pills. Taylor had collated examples of accidental strychnine poisoning, caused by medicine overdoses, or mistakes in dispensing; it was used as a stimulant in cases of paralysis. Strychnine 'has been hitherto considered to be beyond the reach of the common class of criminals,' but had recently been used in some areas for poisoning vermin; 'it may thus find its way into the hands of persons as a substitute for arsenic.' Now that arsenic sales were regulated, poisoners would need an alternative. A surgeon, such as Palmer, would be ideally placed to identify an obscure poison to select as their murder weapon. Taylor only had one example of a trial for murder by strychnine, which had taken place in Canada in 1851.

Had Taylor been furnished with Jones's and Mills's accounts before he had performed his tests, his analysis could have homed in more precisely on strychnine. There were methods for separating it out, but it was complicated, and his analysis had been beset with challenges.

In 1843, he had investigated two separate deaths of children who had died of opium poisoning. He had not been able to detect the poison in their bodies, but every other piece of evidence – their symptoms, the fact that one child had been prescribed opium and that a bottle of it had been found by the bed of the other – showed that they had been administered it, and it was enough to show that it had caused their deaths. But they were accidents; no one would hang.

The public were used to the certainty of the Reinsch test, with Taylor holding aloft copper gauzes dulled with the metallic deposit of arsenic, or a test tube of white crystals of arsenic oxide. There was no such certain test for strychnine, Taylor claimed. When he later wrote the 1859 edition of *On Poisons*, the tests for strychnine he describes require colour and precipitate tests, just as arsenic had before the innovations of Marsh and Reinsch. The presence of organic materials – blood, mucus, partially digested food, faecal matter – interfered with the purity of colour tests.

Other witnesses gave evidence at the inquest – a chemist who said Palmer had bought strychnine from him, and another whom Palmer had approached, asking how much strychnine would kill a dog. Mills was examined again, saying that she had seen Palmer rooting through Cook's pockets, and there were questions over Cook's betting book, which had vanished.

The inquest came to an end on the Saturday. Ward summed up favourably to Palmer, who was still languishing in bed. As objective as Ward may have tried to be, it cannot have been easy to accept his friend's possible guilt. He poured scorn on Taylor; he hadn't found any strychnine, so he couldn't say that it was the cause of death. Ward claimed that Taylor had only hit on it after hearing the words of a chambermaid, as well as 'some other witnesses' – placing no emphasis on the fact that Jones, himself a surgeon, had identified tetanic convulsions. Ward had prepared himself with a copy of Taylor's latest work, and challenged Taylor's own words – that in a recent strychnine poisoning mentioned in the book, the heart and lungs had been full of blood; Cook's had been empty. Palmer had behaved as a friend to Cook, Ward went on, so why would he kill him? And, Ward whined, he objected to the jury forcing him to sum up so hastily.

The jury had made up their minds before Ward even began. They took only six minutes to decide that Cook had 'died of poison, wilfully administered to him by William Palmer'. The police arrested Palmer and waited by his bed. He would also be going to court for forging his mother's signature on some loans. A crowd gathered outside, hoping to catch a glimpse of him, but it was several days before he was deemed well enough to be transported to Stafford gaol.

Beware how you belie the dead

The police were busy with their investigations; on 19 December, Hatton, Chief of Staffordshire Constabulary, went to London to 'prosecute the investigation' further. The following day, Ward had a letter from the Secretary of State's office directing that the bodies of Anne, Palmer's late wife, and Walter, his late brother, be exhumed. The two coffins were removed from the Palmer family vault behind the parish church. Palmer was a regular worshipper at St Augustine's; it stands opposite the red-brick Georgian house where he had been born in 1824. [Plate 16]

His mother was still living there in 1855; his father, a timber merchant and colliery owner, had died 'very suddenly, of apoplexy', when Palmer was only 12. Seven of their children had survived infancy, and each was to inherit £7,000 on reaching 21. It was enough to advance them in the world: William's brothers turned to law, the clergy and the mercantile world. William was initially apprenticed to a wholesale chemist in Liverpool, but was dismissed soon afterwards, apparently embezzling funds; he had already started betting on the horses. In 1841, he returned to Rugeley, and was apprenticed to a local surgeon. There are stories of Palmer's bad behaviour, where he carried on with a girl, and made false entries in his master's books, inventing visits to patients. He fell out with one of the surgeon's assistants, and when the assistant's new boots were found slashed with a penknife, suspicion fell on Palmer.

The surgeon had had enough, so the difficult youth became a walking pupil at Stafford Infirmary, following the medical staff on their rounds. After he administered medicine himself, against regulations, and a student prank went wrong when a man drank himself to death (or was he poisoned?), Palmer was asked to leave. He became a pupil at St Bartholomew's Hospital in London, and was so behind in his medical knowledge that he required the aid of a 'grinder'; yet he still found time for the racetrack, as well as more drink-fuelled japes and larks – and women.

By 1846, he was back in Staffordshire. He was now a surgeon, a professional man, and was introduced to Annie Brookes, a pretty girl with a fortune. He wooed her and married her, although her fortune had been complicated by her illegitimacy; after her father's suicide, the money and properties settled on her were subject to a Chancery dispute. Palmer and Annie were married in 1847, and they had five children; only the first, William Brookes Palmer, survived infancy.

Annie died at the end of September 1854, and Palmer received £13,000 from the life assurances he had recently placed on her. It was a welcome sum because her annuity ceased on her death, and Palmer was in debt. Insuring the life of a family member was not unusual; it was the middle-class version of a burial club. At the inquest into Jane Wooler's death, it was asked if her life had been insured, in case it had served as her husband's motive. The issues surrounding life assurances were on the syllabus of medical jurisprudence courses, because surgeons might be asked to assess someone's health before they were signed up. If Palmer had put aside his student carousing and paid attention to this part of the course, then one might suspect that he had put his lessons to sinister use.

Walter, Palmer's brother, had become a corn factor, but alcoholism and gambling led to bankruptcy and estrangement from his wife, Agnes. By the summer of 1855, he was living in Stafford. Several medical men came forward to swear that Walter, accompanied by his brother, had been examined so that he could be proposed for life assurances. Walter went to Liverpool to visit Agnes; when he said he would take the Pledge and abandon drink, she said that she would take him back if he did so. He told her that Palmer had insured his life, and would advance him money on it, but only gave him a few pounds. He returned to Stafford and continued to drink, and in August he was dead. The medical man attending him decided it was apoplexy brought on by drink. Palmer wanted him buried quickly, so Walter was sealed in three coffins and laid to rest in the family vault.

Palmer visited Agnes to tell her that Walter had died. He then wrote to her, saying that she owed him money based on the sums he had lent his brother. Agnes was furious. 'What right had you to lend your money supposing that I would repay it?' She had never received 'a farthing from him', so why should she pay Walter's debts now that he was gone? She did not believe that Palmer's claims were true, as Walter had told her about the life assurances. She signed off: 'Beware how you belie the dead.'

Palmer tried to claim on the assurances he had been able to place on his brother – not all the companies he had approached had accepted him – but there was suspicion. It might have been stirred by a surgeon called Cornelius Waddell, who had seen Walter when he was suffering from delirium tremens. He was asked to examine Walter for an assurance company, but Waddell wrote to them, 'His life has been rejected in two offices. I am told he drinks. His brother insured his wife's life for many thousands, and after the first payment she died. Be cautious.' The life assurance companies smelt a scam, and sent a former member of the London Metropolitan Police to investigate. This was Inspector Field, the model for Dickens's Inspector Bucket in *Bleak House*. He found that Palmer had even tried to insure the life of one of his grooms, George Bate, claiming that he was a gentleman. [Plates 23, 24] Palmer's insurance claims on his brother were refused, but it had taken the death of Cook for Anne and Walter's deaths to be investigated.

I have known 500 deaths from poisoning

The viscera of Anne and Walter Palmer arrived at Guy's Hospital on Christmas Eve, and were safely locked away until Taylor and Rees began their analysis on 26 December. They found antimony in every part of Anne's body that they analysed, with a tiny amount of arsenic in her stomach; they explained this by some commercial preparations of antimony being impure. Her organs were healthy and there was nothing to account for her death, other than the antimony.

They found no poisons in Walter's viscera, which showed no sign of disease apart, unsurprisingly, from his liver.

Taylor and Rees were back in Rugeley in mid-January for two more inquests, and Taylor was part of the high-level investigation into Rugeley's postmaster, which involved magistrates, two Post Office inspectors, and the Postmaster General.

Eagle-eyed police chief Hatton had seen a letter written by Palmer in the hands of Ward, the coroner. This led Hatton to discover that Palmer had seen the private letter between Taylor and Gardner, about the analysis of Cook's viscera. Taylor had to go into detail about what ink he had written his letter with (he made his own), what sort of envelopes he used, how well he sealed them, and which letterbox he used for posting his letters. Ward was compromised and did his best to avoid being examined, promising to attend enquiry meetings, but never appearing. To encourage the reluctant coroner's participation, the magistrates decided to carry out the investigation in private.

To the journalists, Taylor announced his disappointment that so serious an investigation was hidden from public view. People needed to be able to rely on the Post Office 'as a medium of secret correspondence'. He used the service himself when working on cases, and it involved 'the safety or character of a very large number of persons. ... If he could not feel satisfied that in the Post Office his communications were safe from the perusal of other persons than those to whom they were addressed he should be greatly confounded.' The investigation's outcome would have gladdened Taylor: the postmaster would stand trial.

At the inquest into Anne Palmer's death, Taylor and Rees listened to the evidence of friends and medical attendants who had cared for her just before she died. She had not been well in the months leading up to her death, and in September 1854, had been to Liverpool with her husband's sister. She wore a 'thin white dress' to an entertainment at St George's Hall, and thought she had caught a cold. She had cut her trip short, and arrived home on 20 September. She vomited frequently, and Bamford thought that she had English cholera, which was raging through Liverpool at the time – an odd diagnosis, given that she was constipated, and diarrhoea was its major symptom. Her husband took her food and drink, and gave her Bamford's pills; ample opportunity to poison her. Dr Knight was called on 25 September. He also thought it was English cholera, and prescribed a small amount of prussic acid. He asked her what her symptoms were, but she was so exhausted that she couldn't reply. A few days later, she was dead.

Palmer's solicitor, a Brummie called John Smith, interrupted at every opportunity, picking up on any and every flaw he could find. He was just the sort of tenacious, pedantic legal professional that Taylor found particularly irritating. He came forward with a copy of Taylor's *On Poisons* and read out the section on antimony, claiming that the evidence contradicted Taylor's own printed words. As far as Smith was concerned, newly expert in toxicology thanks to Taylor's book, Anne Palmer had not died of antimony poisoning at all. Dr Knight replied that, 'medically considered', little weight should be given to the words of the woman who had attended Anne and reported her symptoms to the inquest; she was not a professional. And, perhaps most importantly of all, the physician who performed the post-mortem said that he did not think it looked like English cholera.

Now came the evidence of Taylor and Rees. Taylor read through his report of their chemical analysis, and on being questioned by Deane, the solicitor for the assurance companies, Taylor suggested that Anne had been given antimony before Bamford had seen her. Deane had asked if the antimony could have been in Bamford's pills, but Taylor thought not: other, more serious symptoms would have appeared if that had been so.

Smith examined him next. He asked Taylor what the smallest fatal dose of antimony was, and Taylor, indicating his own book in Smith's hands, replied, 'You have it there.' It was ten grains. He admitted that he found the vomiting unusual; when antimony was taken in medicinal doses, based on the amount he had found in the analysis, it did not usually cause vomiting. When it did, there would be ulceration in the stomach, and he had not found any. However, Taylor said, 'I have known 500 cases of death from poisoning, very few of which exhibited the same aspects.' Smith asked if Taylor's results would have been the same if the antimony had been taken three weeks or a month before? No, Taylor replied. Smith wagged Taylor's own book at him again, commenting that Anne's symptoms were not the same as those described in the book. Taylor replied that in the book, he had written about 'the taking of a vital dose'.

'Am I to understand,' Smith asked, 'that there may be the administration of poison in some form or other without any one of those symptoms?'

'Sir,' Taylor bristled, 'what I have described in my book relates to the administration of antimony in acute doses. This is no such case.'

He explained that the doses would have been small, possibly medicinal. Anne was very weak, and it 'would not have been proper' for her to have been given it medicinally. When Rees was examined, he agreed with Taylor: Anne Palmer had died from small doses of antimony.

Taylor and Rees had been in Rugeley for two days, in addition to the time they had spent there in December. Taylor asked if his evidence on Walter Palmer might be taken 'before the hour at which it would be necessary for him to leave Rugeley to catch the express at Stafford, as both he and Dr Rees were absent from London at great sacrifice to themselves.' It was a Saturday; the private Taylor, husband and father, was making a stand.

Dr Knight was re-examined. He said that after hearing Taylor and Rees, his opinion had 'been considerably modified' and 'I am inclined to think I must have been mistaken.' He now believed that when he saw Anne Palmer, she 'was suffering from the effects of antimony'. Several surgeons came forward, who had examined Anne for life assurance policies. Smith objected to almost everything every witness said, even demanding that the reporter from *The Times* be made to give evidence, to explain how Taylor's chemical analysis had come to be printed in full in their issue the day it was laid before the jury. Taylor said that the reporter had been dashing off to catch the train, and he had allowed them to transcribe it as long as it wasn't 'published before I had given it in evidence'. Ward silenced Smith simply by saying, 'I cannot call the reporter of *The Times* as a witness. Call on the next witness ...' *The Times*, in rather awkward phrasing, told its readers that there had been an intermediary, Taylor's embargo having had got lost somewhere along the line.

Despite Smith's best efforts, the jury's verdict was another of wilful murder against William Palmer. They declared that Anne had died of antimony 'designedly administered' to her by her husband. It was time for the inquest into the death of Palmer's brother.

Poisoned by gin?

The inquest into Walter Palmer's death went on for several meetings, dragging on until the end of January. Charles M. Gorway, a caricaturist working for the *Illustrated Times*, captured the scene in the Town Hall at one meeting as Taylor, leaning at a lectern, stands above the assembled meeting, a roll of papers in one hand. [Plate 24] He points, perhaps at someone in the room, or to emphasise his words. His head is slightly bowed; not uncommon for a tall man living in a world designed for shorter people. Ward, the coroner, sits at the far left of the image, at the head of the table, his elbows on the arms of his chair, mouth puckered, as he listens. Almost in the middle of the image is the comparatively dainty figure of Rees. Hatton sits at the foot of Taylor's lectern, scrutinising some papers. In the background, Gorway captured the great crowd that filled the room.

Several medical men came forward to say that they had examined Walter for life assurances; they found him healthy, other than his drinking. The symptoms at his death were agreed on by all the medical professionals as apoplexy, both by those who had seen him at his death, and those who only heard about the symptoms at the inquest. They thought it had been brought on by drinking. But Taylor and Rees had a theory, perhaps suggested by evidence given by a chemist from Wolverhampton who said that, in the month of Walter's death, Palmer had bought prussic acid from them. Around the same time, the 'boots' at an inn in Stafford had been asked by Palmer to look after two bottles for him, and he and the innkeeper had seen Palmer decant some of the contents into a tiny bottle. Could that have been the poison?

It was agreed that apoplexy had killed Walter in half an hour, and Taylor said that prussic acid could explain all the symptoms that were reported. Although prussic acid could kill with great speed, it wasn't unknown for it to take longer. All the witnesses who had attended the death, and who had seen Walter's body immediately afterwards, were asked if they had noticed any particular smell in the room; could they smell the telltale scent of bitter almonds? None could, but Taylor explained that the strong smell of brandy, which Walter had been given by his brother, would have obscured it. To prove his point, Taylor had some prussic acid and brandy with him to demonstrate.

Taylor and Rees would not positively swear that Walter had definitely died of prussic acid poisoning. There was no absolute proof; it was such a volatile substance that it had not been found in their chemical analysis, and the symptoms were not as striking as strychnine. Taylor said, 'The symptoms of apoplexy caused by drink are much the same as those produced by the effects of narcotic poison.' They had not seen any sure sign of disease in Walter's viscera that would explain his death; although his liver was unhealthy, it hadn't been feeble enough to kill him. The brain had been too decomposed for the surgeon who performed the post-mortem to form any clear impression either. It was altogether mysterious. 'He might have died either from apoplexy or from prussic acid,' Rees said, 'but which it was I am not prepared to say.'

The other medical witnesses doubted the prussic acid theory, all focusing on Walter's track record as a drinker. 'He was poisoned by gin,' one of them said. 'The word "intoxicate" means to poison.' Perhaps what swung the opinion of the inquest jury was the arrival of Agnes, that most sympathetic of Victorian figures: the wronged, suffering widow. She 'exhibited so much emotion' that it was agreed 'to spare her the pain of giving her evidence in open court'. The 'lady-like' widow was escorted to a private room and she gave her deposition there; it was perhaps a concession not permitted women further down the social scale.

Ward's summing up was entirely against the idea of poisoning; Taylor and Rees had only said that he 'might have died' from prussic acid. He undermined the evidence of the chemist, and claimed it did not mesh with the timing of the evidence of the 'boots'. Even so, the jury almost unanimously found that Walter Palmer had 'died from the effects of prussic acid' administered by his brother.

Of interest to the public

Almost as soon as Cook's case had been reported, the press had descended on Rugeley, and the public clamoured to know more about what was happening in the otherwise unremarkable Midlands market town. There was the glamour of horse racing; the terrifying idea that a medical professional was using his learning to bump people off; the fear of bankruptcy and debt that loomed over Victorian life; the fascination that such an ordinary looking man could have plumbed such evil depths. The carousel swirled around a central void, the poisoning where no poison was found; it was a story with irresistible draw. It attracted comment everywhere, from the pages of medical journals like *The Lancet*, to mass-market, illustrated newspapers. Tax on newspapers was repealed in 1855, leading to a surge in circulation and the birth of titles such as the *Daily Telegraph*; now that the Crimean War was over, Palmer kept readers enthralled. It has been suggested that a highly coloured portrait of Rugeley that appeared in one of the newspapers came from the pen of Wilkie Collins, who would soon become famous for his Sensation novels, the precursors of crime fiction.

The Lancet remarked on Taylor's evidence at the inquest of Anne Palmer, as reported in *The Times*, asking for his clarification on several points. They devoted one and a half sides to his response. Had he really said that medicinal doses of antimony don't cause vomiting? He replied that in his experience, he had never encountered a case where it had, although he conceded that opinion differed. And did he say that antimony had killed her? Yes, Taylor had; Rees agreed with him, and now two medical professionals who had attended Anne concurred. 'You will excuse me from entering into our reasons for this opinion on the present occasion, as this may form a very fair and proper subject for cross-examination at the trial,' Taylor wrote. He corrected a mistake made by *The Times*, which reported him saying that antimony shouldn't be given during fever, whereas, he said, 'antimonial medicines might be fairly prescribed.' He ended by saying that in his twenty-five years in toxicology, he had 'never met with any cases like these', and that how they affected the accused was 'of minor importance compared with their probable influence on society'. Palmer, faced with the prospect of a public hanging, probably wouldn't agree. But Taylor went on: 'the future security of life in this country' depended on the judge, jury and counsel at the trial.

Meanwhile, Palmer was trying to get his trial moved to London, and his petition was printed in *The Times*. He should have been tried at the Stafford Assizes, but he believed 'that I cannot have a fair and impartial trial' anywhere in the Midlands, because prejudice against him was so great that an unbiased jury would be impossible to find. Smith and Serjeant Wilkins were assembling their own panel of toxicological experts. Taylor 'is the principal witness, and, in order to rebut the evidence given by him, it will be necessary that I should have a sufficient number of scientific persons to give evidence upon my trial, most of whom are resident in London.' Palmer pleaded poverty, that he was relying on friends and relatives to foot the bill, so that the cost of the experts would be reduced if he didn't have the added expense of their travel and accommodation.

It is difficult to feel sorry for Palmer, when it is considered how working-class people accused of poisoning fared. They did not have solicitors to cross-examine witnesses at the inquest, and they certainly couldn't assemble their own team of expert witnesses; some even stood trial without defence counsel. With the trial generating so much publicity – Prince Albert bought one of Palmer's horses when his stables were sold off to cover his debts and legal costs – trying to move the trial was worth the attempt, even though it had never been done before and would require an Act of Parliament. He got his wish: the Central Criminal Court Act 1856 – or Palmer's Act – was passed.

Wilkins was a deliberate choice to lead Palmer's defence, given that he was seen as a medical expert, and had scored an acquittal at the recent Wooler poisoning trial. But the serjeant-at-law with medical knowledge had troubles of his own on the horizon.

Journalist Henry Mayhew was one of many scrambling to keep readers up to date with the Palmer investigations in his weekly paper, the *Illustrated Times*. In January 1856, he published an exposé on life assurance scams, and on 2 February he produced a special edition on 'The Rugeley Poisonings'. He accompanied it with Gorway's depiction of the inquest into Walter Palmer's death, buildings around Rugeley such as the Talbot Arms, and figures such as Palmer, Hatton, the 'boots', the coroner, and Taylor and Rees in a laboratory. His illustrations would later be reproduced in a book of Palmer's trial transcript, published by Ward & Lock, but Gorway's image was slightly edited and retitled, changing it from the inquest into Walter Palmer's death to that of Cook. They sliced the 'boots' off from one side of the image, and hid the name of Walter's surgeon, who was sitting at the table by Taylor's lectern.

Mayhew got a scoop: an interview, no less, with Taylor. When he tried to track him down to 'obtain certainly more ample, and probably more correct, details, than any that our contemporaries have yet published', the eminent analyst was still in Rugeley. Mayhew was told he could see Taylor either the next morning at Guy's or at home in the evening. Mayhew was caught in a dilemma, but finally decided to drop in on Taylor 'during the after-dinner period, which is usually devoted to recreative topics'. How pleased would Taylor have been for Mayhew to invade the sanctuary of his home when he had only just returned from Rugeley?

Mayhew had a letter of introduction from Faraday, so Taylor welcomed him into his home 'as I would Professor Faraday himself'. At this point, things become hazy. Taylor would later claim that Mayhew had misled him. Not revealing that he was from the *Illustrated Times*, Taylor would say that Mayhew had announced himself to be from an assurance company. Mayhew, however, stated that Taylor knew he was a journalist, and consented to their conversation being published, as long as nothing went into print that would 'be prejudicial to the interests of justice'.

Taylor harped his usual refrain, that 'secret poisoning was on the increase' but that the public should not be alarmed: 'the progress of chemical science had done more than keep pace with the progress of the hateful crime.' Would-be poisoners should be deterred by 'the skill of the analyst'. Taylor asked Mayhew to state specifically that 'by analysis the chemist could *almost always* detect' poison, and where it failed, as in the case of Cook, '*physiology and pathology*' would invariably suffice to establish the cause of death.'

In newspaper reports of the investigation into the Rugeley postmaster, it was said that Taylor worked on nearly 150 confidential cases annually. Was this an exaggeration? Taylor explained that he sent about that many letters connected to chemical analyses each year, but only worked on about twenty to twenty-five cases. Mayhew still marvelled at this, calculating that Taylor had worked on 500 analyses in his quarter-century career. Taylor said that, of these, about three fifths would result in detecting poison, or 'there was reason to believe that it had been administered.' This was an important point, considering that nagging gap at the centre of the Palmer case. Arsenic or laudanum were the poisons that he usually came up against, so readers who had been paying attention would realise that strychnine, antimony and prussic acid were unusual murder weapons.

Why did people poison? In most instances, Taylor replied, it was 'with a view to obtaining burial-fees', or to remove an unwanted spouse. But these cases 'occurred almost exclusively among the lower classes.' Without saying so, Taylor implied that a middle-class poisoner was rather different. Had Taylor met with life assurance murders before? Taylor announced the rather surprising belief 'that many persons whose lives were insured were poisoned with spiritous liquors', and said that anyone causing someone to take a fatal quantity of 'intoxicating drink was as much a poisoner as one who administered a fatal dose of poison.' He agreed with the doctors in Rugeley: Walter Palmer had 'been poisoned in the first instance with gin'. Taylor was speaking unguardedly.

Mayhew asked Taylor if he would show them the antimony he had extracted from Anne Palmer. Taylor left the room for a moment, Mayhew entertaining 'unpleasant doubts as to the form in which the antimony might be presented to us'. But when Taylor returned, he had with him a small bottle 'closed with a glass stopper, and containing apparently only a roll of white paper'. He unrolled it and revealed several 'copper plates' – the copper gauzes of the Reinsch test – grey with 'the dark metallic poison precipitated upon them'. Mayhew possibly asked Taylor's opinion on who had given Anne the antimony. Although the question wasn't published, Taylor said, 'it formed no part of his business to say how the poison that was there on the strips of sheet copper before us had been introduced.' All that was asked of him was to explain that he had extracted it from her body, 'and that it was impossible that the substance could have been found in the system unless previously introduced into it during life.'

Taylor didn't name Palmer in the article, but in his opening lines, Mayhew had assumed the man's guilt by declaring, 'Palmer had reckoned upon the impossibility of discovering any traces of poison in the bodies of his supposed victims.' That 'supposed' arrives too late in the sentence for most readers to notice.

Scant justice to the prisoner

In March, Taylor had to go to Stafford, to give evidence at the trial of the Rugeley postmaster. The contents of the letters were read in court, and printed in full in the newspapers. The postmaster was found guilty of having opened the letter, with a recommendation to mercy for his previous good character. He was sent to prison. The level of corruption and bribery this exposed, with Palmer sat at the centre of it, did little more than amplify his guilt. Why else would he have led a good man into an evil path?

At the same Assizes, the grand jury threw out the possibility of Palmer being tried for murdering his brother, the evidence not convincing them that there was a strong enough case, but they passed a true bill for the other two cases. Palmer would be standing trial for murder.

Meanwhile, Taylor had been ducking missiles flung by Smith. At the end of March, *The Times* saw fit to publish some correspondence between Palmer's solicitor and the Home Office. As well as moving Palmer's trial to London, Smith was demanding that Taylor reveal exactly what tests he had used on Cook's remains. He had not been specific at the inquest, and the panel of experts were already assembled, claiming that 'strychnine can and ought to be found, if administered'. They queried the chemical evidence in the case of Anne Palmer, that antimony 'is not eliminated in from fifteen to twenty days'. *The Lancet* also picked up on this, saying that Taylor had implied that the antimony had been administered 'within a few days prior to her death' but the antimony found in her organs could have been there for many months. Yet they were ignoring Taylor's point: it was found in a partly soluble form in the contents of her stomach and intestines, which showed she had taken it just before her death. It was not the same as the antimony found deposited in her other viscera.

Presumably guided by Palmer's expert witnesses, Smith questioned why Taylor and Rees had found a tiny amount of arsenic in her stomach, but nowhere else. 'I submit the chymistry is defective, and therefore not trustworthy, which failed to detect it in other parts.' But this was printed in *The Times*, and Taylor had no way to respond, short of writing a letter in answer, and he had already told *The Lancet* that he wouldn't comment on aspects of the cases that might come up in cross-examination. Smith was undermining Taylor and Rees's analysis in a highly public fashion before they were in the witness box.

Deliberately goading them, Smith was attacking Taylor's expertise in order to force his hand. But the Home Office, guided by Taylor, refused Smith's demands. In provoking Taylor, Smith had probably made him deliberately stubborn and obstructive. *The Lancet* lamented Taylor's secrecy, urging him to give Smith what he wanted. Taylor's insistence on keeping everything that was a matter for cross-examination confidential was 'to leave a question which requires careful examination and deliberate research to the chance information of the witnesses summoned and the skill of the barristers employed. Such a course is not the most conducive to the discovery of truth; it is scant justice to the prisoner.' It raised questions about the way in which scientific evidence was managed in the courtroom, with 'no external observation to inspire the public mind with confidence in the results'. Finally, Taylor agreed, and sent the information that Smith had requested.

Taylor was in a terrible predicament. He may have felt unsure about the strychnine tests himself, hence his prevarication. He had known from Gardner's letter than it was one of several poisons that Palmer might have used, but the condition in which he had received the viscera presented problems for analysis, and the letter may have arrived too late anyway. It was a relatively new poison, and one that Taylor had little or no experience of, aside from a few vivisection experiments in the laboratory. It was terrible luck that the first case in which he encountered strychnine was so high profile, one of the most famous poisoning cases of the Victorian era. But the other experts were just as inexperienced, though they might not admit it. Taylor was attacked for not being public enough, while a postmaster had opened his confidential post, and he had been, so he would claim, tricked into a newspaper interview. He couldn't win.

To make matters worse, he may have discovered who the defence counsel had selected as their panel of experts. Prominent amongst them was Henry Letheby, a name to fill Taylor with dread, recalling their public spat after the Tawell trial. It may not have helped matters that a couple of years earlier, Taylor and Rees had edited Jonathan Pereira's *Elements of Materia Medica and Therapeutics*. Pereira had just died, and Letheby was his former assistant; Letheby may have felt that the inseparable pair from Guy's Hospital had usurped him. Bristol-based chemist William Herapath would also be on the defence team. He was a well-known toxicologist, who had secured a conviction at the first trial for which an exhumation was required for chemical analysis. He covered cases in the south-west of England, examples of which appeared in Taylor's books. Taylor, meanwhile, recommended expert witnesses to back up the prosecution. He included his friend, the chemist Professor Brande, and the famous Robert Christison.

The stage was set for the showdown at the Old Bailey, but who was on trial: Palmer, or toxicology itself?

Hooted and hissed

In April, Taylor was embroiled in another Staffordshire case, when Catherine Ashmall's body was exhumed from a churchyard 8 miles away from Rugeley. Rumour, started by a relative who was perhaps obsessed by other local events, suggested that she had been poisoned by her husband. But the post-mortem showed nothing suspicious, and neither did Taylor's analysis. He did not attend the inquest, and instead sent a report. He was busy preparing for the Palmer trial, and as Ward was presiding over the inquest, he may have wished to avoid the compromised coroner.

Taylor stated that, as Catherine hadn't had a medical attendant during her illness, it was impossible to draw a proper conclusion, but if the servant who reported the symptoms had recalled everything correctly, then there was nothing she had said that would contradict the idea that death had been due to natural causes. He did not claim that someone had been poisoned without good reason.

Catherine's husband was a kind man, and with no motive or evidence that Catherine had been poisoned, the jury returned a verdict of death by natural causes. They received a round of applause. It was very different from the Palmer cases. The originator of the rumour was 'hooted and hissed' as they left the inquest venue, and narrowly avoided 'a ducking in the horse trough by several rough lads of the village'.

The lives of sixteen millions of people

The trial began on Wednesday, 14 May 1856. It being such a high-profile and difficult case, there were three judges, and the prosecution was led by the chief law officer, Attorney General Sir Alexander Cockburn. A huge number of witnesses would be called by both prosecution and defence, and the trial would finally end nearly a fortnight later; most murder trials at this period could be over in a matter of hours. Huge crowds gathered outside the Old Bailey; the case 'had excited such terrible interest in every fireside in England, and … had caused undefined sensations throughout the length and breadth of the land'.

Lord Chief Justice Campbell was the principal judge. Taylor had shared a courtroom with him in 1851, when penniless Sarah Chesham stood trial without the benefit of defence counsel, or her own team of scientific experts. Campbell had sentenced her to hang. He evidently took a dim view of poisoning; when he offered Palmer a chair, rumour went round that he had already decided that the surgeon-cum-gambler was guilty.

Serjeant Wilkins was nowhere to be seen. In ironic echoes of Palmer's financial nightmares, Wilkins had fled to France to escape his creditors. Palmer's defence was led instead by Serjeant Shee.

The prosecution's first witnesses were those who had been around Cook and Palmer in the days leading up to Cook's death, and people who could comment on Palmer's dire financial situation: even the imprisoned postmaster tasted freedom for a day as he gave evidence. There were the men who had performed the post-mortem; it transpired that it had been carried out in Palmer's presence, and he had done what he could to sabotage it. He had proffered brandy to the inexperienced surgeon's apprentice who cut into the body, and he nudged them as they removed the stomach, spilling some of the contents into the body cavity. He then tried to bribe the man who would drive the jar of viscera to the station; it was no wonder that Taylor had received them in such a state.

The expert witnesses, commenting on the scientific and medical evidence, explained cases of tetanus and the symptoms which accompanied them. Among the medically trained male professionals were women who had been at the bedsides of suffering patients. Eminent men such as Sir Benjamin Brodie said they couldn't think of any natural disease that Cook could have been suffering from: it didn't sound like 'traumatic tetanus', caused by a wound; logic would therefore suggest that it must have been the result of strychnine.

Witnesses who had seen cases of strychnine poisoning were examined. Two of the cases were tragic accidents, patients who had died after being given strychnine as medicine by mistake. In one case, the chemist had felt so guilty that he had committed suicide.

Morley, a surgeon from Leeds, had experience of a recent criminal poisoning by strychnine, a copycat case, just as Taylor had feared. Following Taylor's inquest evidence appearing in the newspapers, an unstable man called William Dove assumed that strychnine left no traces, and he used it to murder his wife. Morley had attended Mrs Dove, and described her symptoms to the court. His analysis had proved the presence of strychnine in her body.

Taylor's moment in the witness box came on 19 May. Many laboratory animals died horrible deaths for the sake of this trial, but strychnine was a relatively new poison. In the past, Taylor had been able to examine suicides who had taken arsenic, prussic acid or laudanum, and could study the effects in hospital. Having never seen strychnine in a human being, he could only experiment on animals and imagine how those effects would scale up. In his writing on the Tawell trial, it was clear that Taylor was dubious about vivisection.

Strychnine analysis involved a colour test, and a taste test, looking for bitterness. He told the court that 'these colour tests, as they are called, are, I think, very fallacious.' Having administered strychnine to four rabbits, Taylor and Rees used the colour tests to search for it in their remains. One test yielded a positive colour result, and another resulted in a bitter taste, but no discernible colour change. They did not identify strychnine at all in the final two, either by colour or taste.

Taylor explained that they had been unable to find the strychnine because it had been altered on becoming absorbed in the blood. As far as he was aware, there were no tests available to find strychnine once it had left the stomach contents and had been absorbed, hence he had not found it in Cook's viscera.

He made a point of how difficult his task had been made by the condition in which he had received Cook's remains. 'The part which we had to operate upon was in the most unfavourable condition for finding strychnia if it had been there,' Taylor explained. 'The stomach had been completely cut from end to end; all the contents were gone, and the fine mucous surface, on which any poison if present would be found, was lying in contact with the outside of the intestines, all thrown together.'

He was quizzed about the antimony he had found, but said that he did not think it had caused Cook's death. Then Taylor was passed from the hands of the prosecution to Serjeant Shee. He cannot have been relishing this moment, but he had suffered dissections by defence counsel before.

Shee began by anatomising the meaning of the word 'trace' when used by chemists – his object presumably being, rather like Ballantine at Hannah Southgate's trial, to minimise the evidence by emphasising the very tiny amount of poison found. Did 'trace' mean an imponderable amount? Taylor parried, reflecting Shee's own words back at him: several traces together 'would make a ponderable quantity in the whole'. Shee read out the letter Taylor had written to Gardner, and latched onto the closing words: 'the deceased must have died from the effects of antimony.' Hadn't he just said that the amount of antimony he had found wasn't enough to account for death?

'Perfectly so,' Taylor coolly answered him. But what he and Rees had found would not have been all that Cook had taken. And, at the time of the analysis, without the benefit of a clear account of Cook's symptoms, it was a reasonable conclusion to reach.

Shee asked if quack medicines could explain the antimony that Taylor found. Of course it could. Shee pounced. Considering that the antimony could have come from any source, and considering that Taylor had been consulted by Cook's stepfather, who had suspicions about the death, did Taylor feel himself 'authorised in giving an opinion that [Cook] died from the poison of antimony?' It was an attempt to undermine Taylor's authority, and Taylor's response displays his growing exasperation. But it also shows his own rhetorical ability, highlighting Shee's sleight of hand.

'You are perverting my meaning entirely, I must really say,' Taylor sighed.

Shee dived back into the letter, and Taylor rebuffed him again, so Shee wielded instead Taylor's *On Poisons*. Taylor's own words were being twisted and turned against him, but he had borne the same attack before. Shee, perhaps unprepared after replacing the fugitive Wilkins at the last minute, made the same mistake as Smith, and was quoting at Taylor from a section about large doses of antimony. As Taylor had been dealing with small amounts, he fended him off with ease.

But there was a weakness in Taylor's armoury. Of the few rabbits that Taylor had experimented on with strychnine during his career, he had experimented on five 'when I first began to lecture on medical jurisprudence,' and four following the inquest into Cook's death. To justify this apparently low number of animals that he had subjected to a painful death, Taylor said, 'I have a great objection to destroying life except from absolute necessity. … Every toxicologist cannot sacrifice one hundred rabbits when the facts are well ascertained from other sources.' Shee asked Taylor if he had been rash to 'judge of the effects of strychnia poison on man, by so small an experience as that of ten animals of a particular species,' but Taylor batted him away.

'I think you must add to experimental experience something like a study of poisons for twenty-five years; that is, study on the relations of others, and the collection of cases.'

Shee had Taylor's book in front of him. He had only to leaf through it to see how thoroughly Taylor collated cases from many sources, but it would not suit his purposes to do so. He continued his attack on Taylor's weak flank.

'Would a dog not be much better?' he asked.

'Dogs are very dangerous to handle,' Taylor said in reply. 'I have a very great disinclination to meddle with them.' He explained the differences between dogs, cats and rabbits for the purposes of vivisection, then Shee changed tack, and focused on what Taylor had said at the inquest.

Shee read out the report of the chemical analysis, but he didn't examine Taylor about it. Instead, he appears to have painted Taylor as a callous scientist, with no thought for the fate of the accused.

'Your evidence having been given on that occasion, you returned to town, I suppose?'

Taylor had; he heard of the jury's verdict once he was back in London.

'And you knew, of course, that his life depended in a great degree on your opinion?'

A less experienced witness could have tumbled into this trap, but Taylor was solid in his understanding of what an expert witness did. 'No; my opinion was in reference to death from poison – I expressed no opinion of the prisoner's guilt.' He had abstained 'from all public discussion of the questions which might influence the public mind'.

Oh, really? So why had he written to *The Lancet*? To correct several misstatements made about his evidence. 'And did you think it a right thing to publish that opinion before the man was put upon his trial for his life?'

There had been articles in the press claiming that if strychnia caused death, it would always be found in the body. Taylor said that this was incorrect. Other articles stated that it was impossible to kill with small doses of antimony. Taylor denied this too. If he had left those statements unchallenged, 'there is not a life in this country that would be safe: that is what I say.' If William Dove was anything to go by, Taylor had a point.

Shee accused Taylor of leading Palmer to the scaffold, to which he replied, 'I hope he will be acquitted if that is concerned, but I say that the lives of sixteen millions of people are of greater importance than that of one.' Taylor reiterated that the issue of the antimony was unusual in the Rugeley cases, and that was what he had addressed in *The Lancet*; he had not aired any view on Palmer's guilt or innocence. Besides, there was Smith to deal with who 'has circulated through every paper that Dr Taylor was inaccurate – I had no wish or motive to charge any one with poisoning, but my duty concerns the lives of all, as well as one person.'

But what about the *Illustrated Times*? 'Did you allow a picture of yourself and Dr Rees to be taken for the purpose of publication?'

Shee was referring to the graphic that had appeared in the Rugeley special edition, where it seems that Taylor and Rees had posed for an artist – perhaps even a photographer – in the laboratory. [Plate 21]

'If you will be so good as to call it a caricature.' Taylor, the artist, stepped aside from the toxicologist to comment. No – he had not allowed it, and he had had no idea that it was going to appear. He hadn't even received Mayhew at the laboratory. When Mayhew arrived at Taylor's house with his letter of introduction from Faraday, Taylor claimed that he hadn't realised that he was talking to a journalist. 'On my oath I did not – it was the greatest deception that was ever practised on a scientific man; most disgraceful.' Taylor described his meeting with Mayhew, and that Mayhew had later returned, revealed his identity, and showed him a printer's rough of the interview. Incensed, Taylor had crossed out pieces that referred too closely to the Rugeley cases, but somewhere behind his fury, perhaps he may have been glad of the opportunity to air his views to the public.

The caricature, as Taylor described it, puts Taylor and Rees in a tight spot, but it bears comparison to the other images in the *Illustrated Times's* Rugeley supplement. The caricature is on page 88, and beside it is one of Ward, in profile, seated in a large wooden chair, looking through some papers. [Plate 22] On the facing page is a large caricature of the portly and determined William Palmer, flanked by the 'boots' from the Stafford inn, and George Bate, the groom whose life Palmer had tried to insure. [Plate 23]

Below this triptych, taking up nearly half the page, is Gorway's illustration of the Walter Palmer inquest. [Plate 24] On the far left is Ward, almost identical to his solo portrait on page 88, in profile, in his chair. Rees is sat at the table, facing forwards, as he does in the laboratory image. Taylor is in profile, facing the left, although in the laboratory image he is in profile facing the right. In the far right-hand corner of the image is the 'boots' again, his clothing and pose identical to that in his solo portrait, although he was cut from the image when it was repurposed by Ward & Lock. To the left of Taylor's pointing finger is Bate, another profile portrait, again, in pose and clothing identical to his solo portrait, but flipped from right to left.

It's almost certain that Taylor wasn't lying when he protested that he had not posed for the laboratory image in the *Illustrated Times*. It seems separate caricatures had been worked up based on Gorway's inquest image. In order to supply some lab equipment for Rees and Taylor, the artist had only to refer to a chemistry textbook, or a chemical supplies catalogue. In the inquest image, Taylor is looking down at the audience from a lectern, and in the laboratory image, his head is at the same angle as he looks down at his experiment. His side parting appears on the right in photographs, as it does in the inquest image, which means that it's on the wrong side in the laboratory illustration. It is therefore unlikely to have been taken from life, but copied.

Shee put it to Taylor that he had said, 'He will have strychnia enough before I have done with him.' This was yet more for Taylor to deny, and vehemently so.

'I have never said anything so vulgar or improper that I know of, never to my knowledge.' Taylor knew who to blame for this slur against him. 'It is utterly false, and Mr John Smith, the person who has suggested that to you, has been guilty of other false statements, both in his letter to Sir George Grey, and upon other occasions, he has misrepresented my statements, and evidence, and opinions altogether.'

Having tried to demolish Taylor's character and reputation, framing him as a verbally incontinent liar and a maker of threats, Shee returned to Taylor's practical inexperience of strychnine poisonings. It was true that Taylor had no personal experience of the poison in humans, but he nimbly skipped around Shee's verbal twists and turns.

Shee suggested that Cook beating on the bed was inconsistent with tetanic symptoms, and gave it a name – malleatio – that Taylor later mocked when writing about Cook's death. 'The ringing of a bell might as justly have been transformed into a symptom, under the name of "tintinnaculatio".' Despite Shee's lengthy prodding, and sometimes misleading questioning, Taylor held firm. Based on what he knew, and comparing it to the other cases presented in court by the prosecution, in his opinion, Cook had died from strychnine poisoning.

Rees was examined, and backed up Taylor's opinion. Then William Thomas Brande, professor of chemistry at the Royal Institution, appeared. Taylor had consulted him for his opinion on their antimony test. 'My object was to make an experiment that would satisfy me as to that being a very excellent mode of detecting antimony,' Brande said, 'and I did satisfy myself.' With that, Christison was called.

Having not been pursued through the newspapers by Smith, or caricatured in the *Illustrated Times*, or critiqued by *The Lancet*, Christison could be examined solely on the scientific evidence. He had more personal experience with strychnine than Taylor, even having seen experiments in Paris as a student when it was first identified. He had 'frequently seen experiments tried upon animals': frogs, rabbits, cats, dogs, 'and one wild boar'. During Christison's long examination, he largely agreed with Taylor's points, which had been submerged by Shee's attacks. But even the mighty Christison would not escape from having his own textbooks recited to him, and he was forced to explain certain passages to the court. His examination closed with some questions from Cockburn on the colour tests, Christison stating that they 'are not to be relied upon in the case of strychnia in an impure condition'; they would even prove difficult if the strychnine was pure.

The scientific witnesses paid for by the defence were not going to agree with the prosecution. Letheby made the surprising claim 'that of all poisons, either mineral or vegetable, strychnia is the most easy of detection after death'. But it's clear from his examination, as well as that of Julian Rodgers, whom Taylor had disagreed with at the Dore trial in 1848, that they could only speak of experiments on animals, when they had been searching for a known poison. Herapath could talk of one case where he had identified it in human stomach contents, using three colour tests, but the contents hadn't been scattered through a jar of viscera. Nunneley, who had worked with Morley on the Dove case (the two men were split between prosecution and defence), was examined at length. He was convinced that Cook's symptoms weren't those of strychnine poisoning.

The defence witnesses were all confident that if they had been presented with Cook's empty stomach, they would have been able to find the strychnine. But their conviction sounds hollow when they had never actually been faced with the same situation.

After being carefully examined by Shee, it was up to Cockburn to demolish them.

'Have not you said,' Herapath was asked, 'and said more than once, that you had no doubt that strychnia had been given in this case, but that Professor Taylor had failed to find it?'

Herapath denied it. He even denied the accusation that he had been heard to claim as much within earshot of the Mayor of Bristol. It was an embarrassing slip. If Herapath really did believe that Cook had died of strychnine poisoning, then he must also believe that the man who was paying him was a murderer. But then Herapath changed tack slightly. 'I don't deny that I had a very strong opinion from the newspaper reports.' Others had expressed an opinion; why couldn't he? So Herapath dropped in a mention of the *Illustrated Times*, just to remind everyone of Taylor's own embarrassment.

Herapath's evidence was awkward. He wasn't a medical man, so could only comment on the chemical analysis, and could only say, repeatedly, that if strychnine had been present, then it should have been found. Not that he said so, but Herapath was suggesting two possibilities: either Palmer was innocent, or Taylor was incompetent.

Letheby boldly declared that Cook had not died of strychnine poisoning anyway. The symptoms were all wrong. He couldn't have rung a bell, and his symptoms took too long to come on – at least, so Letheby claimed, based on his animal experiments. But how could he be so sure? Had he ever tried to make an animal ring a bell? Herapath had experimented on about nine animals, but Letheby must have filled every small creature within a 10-mile radius of his laboratory with terror: he had tested strychnine on 'pretty well fifty' animals over the past few years, no less than five in the last two months. He had seen four or five cases of human recovery from strychnine poisoning, one of which had involved Dr Pereira dosing the patient himself.

In cross-examination, Cockburn dredged up Letheby's involvement in the 1850 trial of Ann Merritt, a woman who was accused of poisoning her husband with arsenic. The timing of the poison's administration was crucial, as it would indicate whether it was an accident or murder. Letheby had stated that, from the amount of arsenic he had found in the liver, he could be confident of the time, and that led to Merritt being condemned to death. But several medical men – among them Sir Benjamin Brodie – had said that it was chemically impossible to be so specific. But, Letheby now told the court, it wasn't as bad as it might seem: 'the woman was not pardoned, but transported for life.'

The many other medical and scientific witnesses for the defence had been chosen to bring into doubt the idea that Cook had died of strychnine poisoning. They claimed that a syphilitic sore throat could have been the route through which he was infected by tetanus. They suggested all manner of other causes: a cold, pectoris angina, epilepsy. They did not create a united front, a single theory that would have been more convincing against that of the prosecution. *The Times* commented that their differences meant the jury 'might naturally be led to think that their opinions were not sufficiently authoritative to destroy the testimony of facts and the deductions of common sense.'

The medical and chemical evidence for the trial was of huge importance, but so too were the glimpses of Palmer and his desperate situation that came out in the evidence of the eyewitnesses. Campbell took one and half days to sum up so much disparate evidence, not helped by several interjections from Shee.

Finally, on the twelfth day of the trial, the jury retired to make their deliberations. After nearly an hour and a half, they found William Palmer guilty of murder. In sentencing Palmer to death, Campbell said, 'Whether it is the first and only offence of this sort which you have committed is certainly known only to God and your own conscience.' Palmer would never be tried for the murder of his wife.

A bitter taste

Just over a fortnight after Campbell's sentencing, Palmer was hanged in front of Stafford gaol, before a crowd of thousands. He made no confession before his execution, only commenting that he was 'innocent of poisoning Cook by strychnia'. Was he innocent after all? But he knew that there were efforts to challenge his sentence, so a confession would have rendered his sentence inescapable. If he had used nux vomica or brucia, rather than pure strychnine, his words may even have been a confession in a roundabout way.

Palmer left behind a legacy, his case highlighting problems with the legal system. It helped no one come at the truth if experts like Taylor, Christison, Herapath and Letheby were forced into opposition in court. It would have been better had the four men shared their research, deterring any future William Palmers. Instead, the bitterness engendered by the difficult case, not helped by the sensationalising press, meant that for years afterwards, the English toxicologists took swipes at each other.

Most of the press agreed that the right verdict had been reached, but they expressed a wish for change in the way that medical and scientific evidence were handled in the courtroom. 'Scientific men are retained for or against the person accused. Instead of appearing in the witness-box as the expounders of science, they appear more in the character of advocates,' *The Lancet* said. The expert witnesses selected by the defence did not emerge favourably. 'There were as many opinions as men,' *The Times* remarked.

As debate raged on, Taylor took to the pages of *Guy's Hospital Reports*. 'On Poisoning by Strychnia' addressed the issues surrounding strychnine poisoning and its detection, as well as comments on the trial. Demand was such that the article was published as a separate 152-page pamphlet. It is not a wholly sober, objective work. Its careful tabulation of strychnine cases and Taylor's usual clear explanation of poisoning for the non-expert reader are laced with anger. Its tone is heavily sardonic, breathless sometimes as Taylor races to make his points. His comment about 'tintinnaculatio' appears in a sarcastic footnote. Being openly rude was not unusual in journal articles at the time, but page after page of defensive, enraged fury is astonishing to read.

As if trying to remind everyone that his expertise did not begin and end with poisons and Palmer, immediately following the article in *Guy's Hospital Reports* was a short piece by Taylor, analysing water from Iceland's Great Geyser. Unfortunately, it may only have reminded readers of Taylor's volcanic rage.

Mystery would always cling to the death of Anne Palmer, and Taylor's theory of antimony poisoning in small doses was openly debated. In 1857, Taylor wrote another *Guy's Hospital Reports* article, justifying his conclusion as to how small doses of antimony could kill. Since the Palmer trial, there had been two trials for murder by antimonial poisoning, and one man who had been found guilty had confessed to poisoning his wife. It was perhaps the second Palmer copycat murder.

And then there was a misprint in *The Lancet*. A man had committed suicide on the Isle of Wight by taking strychnine, and the medical attendant, Wilkins, had sent the viscera for analysis to a friend, who had passed it on to Taylor. Seeing the chance to clear his name, Taylor had sent parts of the well-travelled viscera to Christison in Edinburgh, and Geoghegan and Maclagan in Dublin, and they all performed the same analysis, the Stas test. Taylor told Wilkins that it 'has been found quite adequate to detect strychnia in the bodies of animals poisoned by a quarter of a grain.' Taylor devoted three weeks to his analysis, and only found a mere whisker of strychnine; the colour changed slightly, but the bitter taste wasn't quite right. 'When a man's life depended on the answer, I should decline to say that such a result was conclusive,' Taylor said.

Wilkins shared Taylor's results with *The Lancet*, quoting the eminent analyst's announcement that if strychnine had been absorbed in the body, it was untraceable by the apparently reliable Stas test. Unfortunately, a misprint occurred, which made Taylor's grand experiment sound utterly bizarre: they had him say that the Stas test was 'quite inadequate to detect strychnia'. Rodgers wrote in, along with Girdwood, assistant surgeon of the Grenadier Guards, to question what the heck Taylor was up to. Why was he expending all this effort on a test that he had declared wouldn't work? Wilkins awkwardly pointed out the error, but Rodgers and Girdwood didn't back down; 'We should not have commented so severely upon the word inadequate, had we not known that Stas's process is really inadequate.' Their own strychnine test, they declared, was superb.

Rather than enter into a lengthy rhetorical tennis match in the pages of *The Lancet*, Taylor wrote about the Isle of Wight suicide and the subsequent analyses in *Guy's Hospital Reports*. Inevitably, it included a dig against Rodgers's testimony at the Palmer trial. Taylor stuck to his claims and apparently made no attempt to try Rodgers's test. Letheby launched himself into the fray, his cage rattled as he had been mentioned in the article. In two letters to *The Lancet*, he criticised Taylor's analytical ability and bragged about his own. Since the Palmer trial, he had had a couple of suspected strychnine poisonings referred to him. He had found strychnine in the analyses, and, he claimed, with the viscera in a worst state than Cook's. He concluded by saying that Taylor's analysis of Cook's viscera had been flawed, yet the colour change Taylor had reported actually showed the presence of strychnine after all. But it had been too slight for Taylor to have placed any strong conclusion on. 'Dr Taylor had not the skill to find it,' Letheby crowed.

And pass away

The *Illustrated Times's* Rugeley special included a depiction of Cook's resting place by the churchyard wall beside two young trees, a heap of earth above him. [Plate 27] Rugeley attracted a pageant of visitors, and the reverend incumbent paid for a stone to be laid over his churchyard's most famous resident. The trees are much taller now, and most of the gravestone's inscriptions have worn away. But surviving on the side of the tomb are two lines from the Old Testament Book of Proverbs:

Enter not into the path of the wicked. Avoid it, pass not by it, turn from it, and pass away.

Chapter 13

Truth Will Always Go the Farthest
1856–57

A very clumsy murder

Although Taylor's professional reputation had come under attack in the aftermath of the Palmer trial, he was still a respected analyst. In Wakefield, Yorkshire, Mr Jackson of Fleet Mills had been accused by the local sanitary inspector of adulterating his flour. When the local chemists could not reach a consensus, Taylor was given the deciding vote. His analysis found that the flour was pure and fit for human consumption, and saved Mr Jackson's business.

The lives of many were saved when Taylor and two other chemists were called upon to analyse tinfoil. It was used to package Adnam's groats, a food for children and invalids. Taylor had apparently analysed the groats before, so that Adnam could boast of the purity of his patent food in newspaper adverts. It wasn't unusual for Taylor's name to appear beside adverts for various foodstuffs, his name a badge of trust. Adnam was claiming damages from the manufacturer of 'Bett's patent metal', which he had been using as packaging, because it had corroded. The chemists found that the foil was only twenty to thirty parts tin; it was mainly lead, and had impregnated the food. Astonishingly, Betts was not sanctioned for selling such dangerous foil for food use, and Adnam did not win his case. It was ruled that he should have realised something was amiss because the foil was so cheap.

But there were always murders to investigate. 'The Bacon Tragedies' began in Woolworth, south London, at the end of December 1856, when two children were discovered with their throats cut. Their mother, Martha Bacon, claimed that an intruder had broken in through a window and had attacked her as well. But a sharp-eyed police constable noticed that the window was fastened from the inside and that the dust on the sill hadn't been disturbed. There were no footprints in the soft soil outside, and there was blood on two knives in the kitchen. Suspicion fell on Martha; blood was found under her fingernails, and once she had been arrested, the matron of the gaol found blood on her chemise. A cut on Martha's neck was examined by medical professionals who decided it was probably self-inflicted, and they believed her to be insane. But Martha claimed that her husband, a whitesmith called Thomas Fuller Bacon, had murdered their children.

He was a violent man, given to drinking, and it was said that Martha's insanity had been caused by him throwing her down the stairs and stamping on her head. Bacon had been on trial earlier that year for arson, setting fire to his home in Lincolnshire. The house and the lives of Bacon's two children inside it were insured, but the judge decided that as the house was worth more than he would have received from an insurance payout, it would have made no sense for him to have burnt it down. Bacon was acquitted, but such was his reputation that it seemed very likely that he was involved with the Woolworth murders. He said it was a terrible thing for his wife to accuse him, and that 'truth will always go the farthest'. Yet there were doubts: although there was a cut on his finger, and there was blood on his trousers, had he been wearing those trousers on the day in question?

In February 1857, while Bacon and his wife were awaiting trial, Thomas's widowed mother was exhumed from the churchyard in Lincolnshire, where she had been buried in May 1855. Standing by the graveside as the coffin was raised up from the earth was the imposing figure of Alfred Swaine Taylor.

It was unusual for Taylor to be present at an exhumation, but as the case was threatening to become as infamous as Palmer's, it's possible that he wanted to avoid the mishaps that had occurred in Rugeley. A local surgeon performed the post-mortem and found that most of the body was decomposed, except the abdomen, which was unusually well preserved. It had been suspected that Ann Bacon had been poisoned, and at the inquest, the surgeon remarked that if Taylor found arsenic, it would 'show that bodies consigned to the tomb poisoned by the action of this pernicious drug will for years retain the appearances peculiar to its specific action as a preservative'. Taylor packed a selection of body parts into jars and headed back to London.

He examined blood taken from Ann's chest, her viscera, and mould from her grave. The blood and the mould did not contain any poisons, but Taylor found arsenic in most of her viscera, as well as antimony and mercury in her intestines. The organs, although shrunken, were otherwise well preserved. Taylor went back to Lincolnshire for the resumed inquest: the arsenic must have been administered during life, but the amount found was tiny, only three quarters of a grain. He stated that, 'whether the deceased died from the effects of the arsenic or natural disease, can only be determined by the symptoms from which she suffered in the illness preceding her death.' He would wait to hear the evidence of the medical attendants.

The local surgeon who visited Ann said, 'she was attacked with what he believed to be English cholera.' The coroner asked if, following Taylor's evidence, he now thought that her symptoms were those of poisoning; yes, he did. The illness had come on suddenly, and Ann had told a local woman that she had fallen ill after eating some broth. There was evidence from a chemist that Bacon had bought an ounce of arsenic only a few days before Ann's death – the poisons register proved it. He had first of all tried to send his employee, claiming it was needed for metalwork, and then went himself, saying it was for killing rats. Just before Ann's death, Bacon had said he had his mother's rent book, remarking to a relative, 'I suppose you know that by my father's will this house, and all that is in it belongs to me and my sister.' There was also a house worth £90 a year in rent, so he wasn't short of financial motive.

The coroner asked Taylor if he had formed an opinion on the cause of death. Taylor said that it was consistent with death from arsenic poisoning. Unsurprisingly, the inquest jury returned a verdict of wilful murder against Thomas Fuller Bacon.

But first he and his wife appeared in the dock at the Old Bailey for the murder of their children. The newspapers explicitly compared the sensation surrounding their trial to Palmer's the year before, 'one that caused an amount of interest and excitement, both in and out of the metropolis, scarcely, if ever, equalled in intensity'. After the first day's proceedings, Martha sat in her cell and confessed to the murders, saying that she had done it alone; her husband had not been involved. When the judges returned the following morning, they had to proceed with the trial, but emphasised the lack of evidence against Bacon, and summed up in his favour. It led to his acquittal; Martha Bacon was found not guilty by reason of insanity.

Bacon had now been acquitted twice. Would he be so lucky a third time? At a pre-trial examination before magistrates in Lincolnshire, he was dressed respectably in a black suit, and was 'perfectly cool and collected', but his temper was ignited when he 'perceived a professional gentleman with pencil and paper apparently engaged in sketching his portrait'. Bacon demanded to know if he had been brought to a court of justice, 'or to be exhibited in my native town?' The magistrates told him that they couldn't prevent the caricaturist, and Taylor, who was present to repeat his scientific evidence, must have wondered if the caricaturist was intending to direct their pencil in his direction.

Bacon's third trial took place in July 1857. There were two bills against him: murder, and administering poison with intent to murder. Taylor showed the court some of the arsenic he had found, both in crystal form, and deposited in metallic form on copper. Having heard the evidence of the surgeon, who said he had prescribed Ann Bacon an antimonial medicine, Taylor suggested that this accounted for the antimony he had found in his analysis. Taylor was asked by the prosecution about 'homicidal mania', presumably in connection with Martha Bacon, which Taylor said 'operates at intervals'. Did it include a penchant for arson? No, Taylor said, but some people claimed it did so.

The defence argued that Bacon's motive was inadequate, and that the evidence failed to show how Bacon had administered the poison. Bacon had rushed at once for medical help when his mother fell ill, and had shown great filial affection. They strongly suggested that it was Martha Bacon who had killed Ann. The jury still returned a guilty verdict, and Bacon was sentenced to death.

He was 'convicted at last', *The Times* remarked, but they were uncomfortable with the sentence. The most damning evidence, they believed, was his purchase of arsenic a few days before his mother's death, and for apparently spurious purposes – but why had he bought his murder weapon so publicly? 'If he was a murderer, then, in act or intention, he was a very clumsy one.' Like the prosecution, their eye had fallen on another suspect: Bacon's wife.

In the end, Bacon was not executed; his sentence was commuted to penal servitude for life. Whatever Taylor felt about the sentence, he was rather pleased with himself: he had managed to find a tiny amount of arsenic in a corpse that had been buried for nearly two years. After doubts over his abilities following the Palmer case, here was proof that the skilled analyst was a force for poisoners to reckon with. But his next case to capture the attention of the public didn't involve poisons at all.

The Waterloo Bridge Mystery

Just as midnight struck on 8 October 1857, a woman walked onto Waterloo Bridge. She came from the Middlesex side, carrying a heavy carpet bag and a large brown paper parcel. She was grey haired, 'of short stature and rather stout', dressed in a black silk mantle and a satin bonnet. As the turnstile was rather narrow, the money-taker, a man called Etherington, lifted the carpet bag over for her. He had been a constable in the Metropolitan Police; he was covering Westminster; he was a man who was 'remarkable for his shrewdness and caution'. He noticed how heavy the bag was, and in the strong light from the gas lamp, he saw an unusual flower design worked at the centre of its fabric. But he didn't watch the woman's progress across the bridge. He thought no more about her until a horrible discovery was made the following day.

Five or six hours after the woman and her carpet bag had come through the toll gate, two lads were rowing up the Thames. In the weak early morning light, they saw a bundle on a buttress under Waterloo Bridge. They rowed nearer and realised it was a carpet bag, wrapped in a cord, with a long piece of string tied to it. They manoeuvred the carpet bag into their boat and, thinking they had found something valuable, went to find one of their brothers aboard his barge on the river.

They were met with a gruesome surprise when they discovered the bag contained human bones and flesh, rolled up in blood-soaked clothes. They reported their find to a police constable, who took them to Bow Street station, and the divisional surgeon, Mr Paynter, was summoned to examine their find. The surgeon fitted the bones together as best he could, and discovered that they formed a near-complete skeleton, missing the head, hands, and feet. He estimated that the body had stood 5′ 8″ tall. Although mutilated, there was enough flesh left to show that it had once been a man, which was also suggested by the short, dark hair on the remaining skin. There was a deep gash in the breast, a stab wound received in life, which matched up with one of several holes in the clothes. The body had been severed by an unskilled hand, using a fine saw, and had been soaked in brine for some days after having been boiled. The clothes were good quality, leading the press to describe the unknown victim as a gentleman. The pockets had all been turned inside out, which suggested that the mystery person had been robbed. Paynter found a dark whisker adhering to the remains, as well as a long, fine hair, which apparently belonged to a woman.

The police circulated Etherington's description of the woman, though it hardly narrowed the field of suspects, especially as Etherington thought her grey hair was a disguise, the result of powder. The paper parcel she had been carrying had presumably contained the body parts missing from the carpet bag, but it hadn't landed on the buttress and must have 'floated away with the tide'. The woman had probably corded the bag once she was on the bridge, so as not to arouse the suspicions of the toll-bridge staff, and used the rope to lower the bag into the river so that it wouldn't make a splash and attract attention.

As crowds rushed to the bridge to gawp at the site of the discovery, and loitered outside Bow Street police station for news, the police realised there were three lines of enquiry. There was an almost full set of clothes, there was the chance that the owner of the bag might be traced, and there must have been a cabman who had taken the woman with her heavy burden to the bridge.

Identifying the clothes was difficult as they had been in brine with the body, and were shrunken and bloodstained. Experienced tailors were summoned and it was their 'decided opinion that they are of foreign make'. Eight or nine gentlemen visited the police station over the weekend to look at the clothes, 'searching for some clue to the discovery of persons who, within the last month, without any known reason, have disappeared, and never since been heard of.' None of them were able to identify the clothing.

Despite so little remaining of the body's flesh, valuable forensic clues were left in the clothes, if they were assumed to have belonged to the deceased. There were three stabs in the back between the shoulder blades, three in the abdomen, and seven over the heart. The stabs in the back and abdomen were quite far apart, which indicated a struggle, but the seven over the heart were close together, which suggested they had finished the victim off. The back of the greatcoat was marked with white, 'as if the murdered man had struggled hard with his back against a white or lightly papered wall.' As the front of the clothing was drenched in blood, it seemed that the victim had lain on his front for several hours after the attack. The clothes had been sliced open up the back, implying that the body had been quite stiff at the time that it was stripped.

The carpet bag was new but old-fashioned, its lining ripped out, with '4s & 8d' in ink on the handle. It was so unusual that certainly anyone who had seen it would remember it – even if they weren't retired police constables – but no one came forward. The police questioned every cabby they could find, but none remembered the woman with her heavy load.

At the inquest, Paynter was asked if the remains could have been used for 'anatomical experimentation'. Paynter refuted this: 'A medical man must have wanted a body either for the muscles, nerves, arteries, or bones. The muscles, nerves, and arteries I can most positively assert have not been dissected, and the bones are destroyed.'

Another clue came to light; the socks of the deceased were found to be 'cotton-ribbed in a very peculiar manner', such as were only made in Germany (in the 1850s, 'Germany' referred to a geographic area in mainland Europe, not to a specific country). The make of the shirt showed that 'he must have worn his shirt collar turned down over the necktie, which again confirms the suspicion that he was not a native of this country.' It would not be until later in the nineteenth century that British men would wear turned-down collars.

Hundreds came to look at the clothes, the police hoping that although most were there to satisfy their morbid curiosity, amongst them would be someone hoping to trace a missing friend or relative. *The Times* declared that it was pointless while the clothes remained in their current state; they suggested that the clothes should be washed and mended, and placed 'upon a lay figure like those which stand in the shop windows of so many tailors'. A reward of £300 was put up by the government for information.

As the area near the bridge was renowned for its brothels, a Reverend Spurgeon issued a broadside called 'Murder Will Out', complete with a garish illustration of a man having his throat cut. [Plate 29] Its prurient details told all about a brothel that had its own torture chamber featuring 'The Man-Cog'. The broadside said far more about Spurgeon than it did about the murder itself.

The superintendent of police at Stafford wrote in with a lead. A man fitting the description of the victim had arrived there from Australia, and had purchased a large number of boots and shoes. He had been carrying a carpet bag like the one described in reports, and it had contained nearly £4,000 in gold. He even wore his shirt collar turned down. His next stop was to be a shoe manufacturer in Clerkenwell, London. Could the mystery be solved at last?

No. After transacting his business, the man had gone to Glasgow, and a more thorough description of his clothes showed that they differed from those found under the bridge. Accusation of 'apathy or negligence in the conduct of their investigations' had been flung at the police, but they had followed up every possible clue that came into their hands and had come up with nothing. It was time to call on Professor Taylor.

Assisted by Paynter, Taylor made his examination over three days. They worked in the presence of Inspector Durkin, the detective who had been on the case since the grisly discovery was made. Taylor was able to add precision to Paynter's initial observations. There were twenty-three parts of the body, altogether weighing 18lbs, about one eighth of the weight of an average adult. The man was aged between 30 and 40, and showed no sign of disease, or any physiological peculiarity that might have helped identification. The body had been cut and sawn while it was in a rigid state. The stab wound appeared to have been inflicted in life, or very soon after death; it would 'have produced rapid, if not immediate death'. But Taylor cautioned that, with the absence of the head and the viscera, they could only speculate as to the cause of death.

He scotched all rumours that the body had been used for anatomy. Taylor announced that he had spent seven years studying anatomy by dissection, and so spoke from a position of considerable experience. 'The joints had been sawn through, evidently with great trouble, at points where a scalpel in the hands of even a young anatomist would have speedily effected a separation of the limbs.' The muscles, blood vessels, and nerves had been cut through 'in all directions'. Neither were there duplicate parts, as 'are commonly found when bodies have been used for the lawful purpose of dissection'. This might not have given the public a particularly comforting view of anatomy teaching.

Taylor performed a chemical analysis on the few fleshy parts that remained. Not to check for poison; without the viscera, there was little point. He was looking instead for any preservative substances, but all he found was salt. The appearance of the flesh and the condition of some of the marrow showed that the remains had been boiled. This highly unusual and rather gruesome aspect of the case was explained by Taylor quite simply. It had been done by someone who had wanted to prevent putrefaction and disguise any unpleasant smells. It was almost impossible to form a reliable idea of the time of death, but deep in the flesh, and in the hip joint, Taylor found putrefaction, which allowed him to date the death to late September or early October.

Taylor examined the clothes at Bow Street police station. He realised that the blood that had pooled at the front of the clothes was concentrated on the left, which suggested it had come from the stab wound over the heart. He warned that there was no direct proof that the clothes had belonged to the body, only that their condition was consistent with the idea that they had been worn by the deceased, and that whoever had worn the clothes had 'sustained serious personal injuries'.

The coroner was taken aback by Taylor's thoroughness, and thanked him for his 'very lucid' report. He and his jury had no further questions, and Inspector Durkin had no more evidence, so it was left to the jury to come to a verdict. They found that an adult male had been wilfully murdered by a person or persons unknown.

Taylor 'took occasion' to thank Paynter for his valuable assistance. Paynter then asked what would happen to the remains. They were in preserving fluid at Bow Street police station, and Inspector Durkin planned to keep them in case any further body parts were found.

The mystery stayed afloat in the public mind. In November, a clerk from the War Office, 'a tall young man with a Glengarry cap', had been arrested for stealing a door knocker on a drunken jape. He had panicked on his arrest as, for reasons best known only to himself, he had some bones in his pocket 'and he was rather afraid that he should be kept in custody for the Waterloo Bridge tragedy.' Just after Christmas, sensation rocked the garrison town of Colchester, when rumour got about that one of the soldiers had confessed to the murder. He had deserted from his regiment and on being brought back to camp, had tried to strangle himself, and then made his confession. He retracted it, saying he knew nothing more about it than what he had read in the newspapers, but not before the town's mayor had become involved. The mayor was a physician and 'expressed his belief that the man's mind was deranged.'

The crime was never solved. The woman on the bridge was never traced, and the victim and his murderer were never identified. But Taylor added it to the annals of medical jurisprudence, including information that the newspapers had left out. The police had continued their investigations, and 'there was reason to believe that the remains were those of a Swedish sailor from a vessel then in the river,' but they had never got any closer to solving the mystery. Complete with Taylor's diagram of the partial skeleton, the case appeared as late as the 1948 edition of *Taylor's Principles and Practice of Medical Jurisprudence*, just as he had originally written it nearly a century before.

Chapter 14

Grieved Beyond all Endurance
1857–59

You did not see me do it

Att the end of 1857, Taylor's bloodstain analysis skills were called on again. John Starkins, a police officer in the Hertfordshire Constabulary, had been found stabbed to death in a pond just outside Stevenage. The prime suspect was one Jeremiah Carpenter, a powerfully built labourer whose employer had summoned Starkins, fearing a robbery. Various clothes and items from Carpenter's cottage bearing suspicious red marks were sent to Taylor to analyse. Taylor went to the inquest and reported that he was able to positively identify blood on Carpenter's gaiters, pocketknife and basket. It seemed that his smock had been scrubbed, and some force had pulled apart the seams, indicating a struggle. The blood was fresh, although Taylor could not say exactly when it had been shed. Neither could he positively say whether it was human or animal, but he could say it must have been shed by something while alive, or recently dead.

There were no eyewitnesses. Carpenter himself said, 'You did not see me do it, nor nobody else.' When his case came to trial in March 1858, he was found not guilty by a jury who said there was strong suspicion against him, but not enough evidence. The local newspaper lamented the evils of the death penalty when a jury wouldn't send a guilty man to the gallows: 'the woodstealer and the pickpocket are shut up in prison, while CAIN goes at large.'

Next, Taylor was involved in another case of a surgeon gone bad. Thomas Monk had occupied several respected positions in Preston, Lancashire, even serving as mayor. But when Mr Turner, one of his patients, died, people were surprised to find that his will was made out in favour of Monk, and that Monk had already walked off with his possessions. Turner had died of a bowel complaint; could Monk have forged the will and poisoned him?

Taylor had said before that toxicology could defend the innocent, as well as uncover crime. In this case, Monk was sentenced to life in prison for forging his patient's will, but Taylor could find no evidence of Turner having been poisoned. 'Only a few small portions of mercury were discovered,' but they could have been medicinally prescribed.

Deadly wallpaper

Public health was consuming Taylor's attention. Two children had died in Romsey, Hampshire, apparently of poison, but Taylor believed that it was an environmental 'local poisonous influence', rather than the hand of a murderer. He reported on the water supply in Whitehaven, Cumbria, finding that water was reacting with iron in the pipes to create carbuncles of peroxide of iron. But he sent huge numbers of people into a flap, including the House of Lords, when he made an announcement that should have shocked no one. It was in 1858, during consultations for the Sale of Poisons Bill, that Taylor pointed out that using arsenical dyes for wallpaper was dangerous.

While explaining that arsenic was used in several industries, Taylor said that 'the largest quantity of arsenic used in this country' was in wallpaper manufacturing. He believed that it was dangerous for people living in houses that were papered with it, as well as for the workers in factories that made it. The pigment, used for a variety of colours, not only green, was almost 50 per cent arsenic, and was loosely put on; it would easily unbind itself and affect residents. This particularly alarmed the government, because at the time, the Inland Revenue's new offices were being decorated with arsenical wallpapers. Taylor had submitted an article by Dr Halley of Harley Street who had suffered from a perpetual headache, dry throat and tongue, and 'internal irritation'. After three weeks, he was completely prostrated and was developing paralysis. Halley performed a test, using paper soaked in ammoniacal silver nitrate, and believed that it proved there was arsenic in the air of his room. He took Taylor a sample of his wallpaper, and Taylor found that the pigment contained arsenic. Halley rid himself of his green wallpaper, and his health returned.

Taylor and Halley were undermined at every turn. The *Journal of the Society of Arts* wrote in sardonic tones that, since Taylor's bombshell, 'we have felt very uncomfortable … when sitting in any room covered with pretty green paper.' They joked that the staff of the Inland Revenue were dedicated martyrs for braving their green walls. Another chemist was commissioned to perform some tests, to verify Taylor's claims. Phillips papered two small closets with arsenical wallpaper, leaving them for seventy-two hours, with a gas light burning for forty-five hours. He found nothing, so declared Halley's test to be bad chemistry. Phillips said that he had arsenical wallpaper himself, and his family experienced no ill effects. Meanwhile, the health of the wallpaper manufacturers went ignored.

Unbowed, Taylor sent the *Journal of the Society of Arts* the section on arsenical wallpaper from his forthcoming edition of *On Poisons*, with added comments. He piled on example after example, culled from a decade's worth of scientific and medical journals, showing the perils of arsenical pigments in paints and wallpaper.

A 3-year-old had displayed severe symptoms of arsenic poisoning after sucking slips of green paper; a baker's loaves, placed hot out of the oven onto freshly painted green shelves, tested positive for arsenic; there were several examples of manufacturers who worked with arsenical pigments falling ill. Taylor suggested that the mystery of the Arzone family, who had all died one after the other in a short period, was explained by arsenic poisoning, thanks to the father's job as a pigment manufacturer. There was a report of green oil paint being used in a damp room, which had led to a 'putrescent and highly disagreeable odour' and the occupants of the room feeling ill. The paint was removed and the smell and the symptoms vanished. A Birmingham physician had fallen ill after redecorating his rooms; he attributed it to the arsenic in the pigment. During the summer, Taylor had wetted some arsenical wallpaper and it had killed some flies.

Taylor blamed Halley's experience, and those similar to it, on arsenical dust; Phillips said his wallpaper was glazed, and Taylor said this locked the dust in. Anyone with unglazed arsenical paper could easy knock the dust off, even in the everyday 'mechanical dusting of domestics'; his assumption was that killer wallpapers would not hang in the poorest homes. An objection had been made to the restriction of the sale of arsenic by a chemist who said that anyone who wanted to poison someone had only to scrape some pigment off their wallpaper.

In Prussia, meanwhile, arsenical pigments had been banned outright. Taylor suggested that Britain should follow suit, and that other dyes should be used, even if it was more expensive. His tone was clear and measured until he reached the end of the article. His trademark sarcasm could not be restrained in the face of such dangerous stubbornness, so he unleashed an acerbic ending, worthy of a *Punch* cartoon. 'If the manufacture of arsenical papers is to be continued, the Prussian poison-symbol of a skull and cross-bones, with the motto *memento mori*, should be printed as a pattern upon it.' It is a tragedy that Taylor did not consider a career in interior design. Meanwhile, William Morris, whose famous wallpapers did not feature skulls or crossbones, used arsenical pigments into the 1880s.

Still intent on proving his point, Taylor wrote to the *Journal* again. A friend of his had suffered from severe eye inflammation after papering their library walls in arsenical wallpaper. They removed the paper, and the problem fled, but the issue returned after dusting some books. Taylor carefully removed some of the dust and analysed it: under a microscope, it was green, and the Reinsch test proved that it contained arsenic. He then took some dust samples from the tops of the instrument cases of an optician's near London Bridge, just around the corner from Guy's Hospital. They had unglazed arsenical wallpaper, and it certainly didn't surprise Taylor to find arsenic in the dust.

The years went by, and more cases came to light of people suffering thanks to their wallpaper, but arsenical pigments were never banned in Britain. Finally, it was consumer pressure, encouraged by the revelations of science and medicine, that led it its use being phased out. Although Taylor had shown that there was arsenic in the dust, and that he believed this was how it poisoned people, it wasn't until the 1890s that the mechanism was finally understood. Fungus in wallpaper paste combined with arsenic in the wallpaper dye to create a poisonous vapour.

Broadly put forth

By the late 1850s, Taylor was a man of great scientific renown. He was a member of the Surgical Instrument Committee, which promoted improvement in their production, and to recommend reward 'for successful invention'. In February 1859, he was awarded the Swiney Prize for his *Manual of Medical Jurisprudence*, at a joint meeting of the Royal College of Physicians and the Society of Arts. This was a silver cup worth £100, containing £100 in gold; it was given every five years for a work on medical jurisprudence.

Many children had been named after Nelson and Wellington. In the summer of 1857, a mineral tooth manufacturer called William Thomas Taylor, who lived in south London, named his second son Alfred Swaine Taylor. The denture maker does not appear to have been a close relative of the medical jurist; neither does he seem to have had any family with the surname Swaine to otherwise explain the choice of name. Unless the two men were acquaintances, it would seem that it was the eminent analyst's reputation that had apparently led to the boy being named after him.

On Poisons went into its second edition in early 1859. It is an exhaustive work of nearly 800 pages, featuring Taylor's own experiences in laboratory and courtroom, as well as examples taken from his wide reading. Updated from the 1848 edition, it made frequent reference to the Palmer trial, with numerous digs at the chemists who had dared to find fault with him. This did not go unnoticed by reviewers, who remarked on his frequent 'controversial and personal tone'. His attacks on 'certain well-known professional men' were perpetual and 'broadly put forth'. Even so, *The American Law Register* felt it was essential reading, taking 'its place among the works always cited in capital trials for poisoning'.

Taylor managed to make jibes at his rivals at the Palmer trial even in passages that had nothing to do with the Rugeley rogue or strychnine. He took great delight in relating Herapath's failure to find anything during a suspected phosphorus poisoning case. The suspect had bought phosphorus under false pretences, there was evidence of him having repeatedly spread it on the victim's bread, the victim's viscera was inflamed and blistered, but Herapath could not find any phosphorous. He argued that it had somehow decomposed and turned into other acids. Taylor wrote that the acids aren't volatile, so should have remained, and would have been detectable 'by a peculiar odour'. What gleeful Schadenfreude Taylor exhibits, echoing the Palmer defence's own words back at them: 'if phosphorous was there, Mr Herapath ought to have found it.'

A high perch can be occupied provided one does not make mistakes. Taylor's reputation had been shaken by the wiliness of a gambling surgeon; it would be shaken again by a bigamous physician.

Some influence or terror

On Sunday, 1 May 1859, Thomas Buzzard called on Taylor at home with two bottles, hoping he might be able to save the life of a woman who he believed was being poisoned. Such a request had precedent: Taylor's intervention had saved the life of Mary Ann Geering's son. 'I do not make analyses on a Sunday,' Taylor later said, but as it involved the safety of a living person, 'I thought it proper to come to a conclusion as soon as I could.' He checked the purity of his hydrochloric acid, and tested the stool sample he had been sent.

He noticed a metallic deposit on the copper, signalling the presence of arsenic, antimony or mercury. Using further tests on the metallic deposit to narrow it down, Taylor isolated it to arsenic. He would be sent three samples in all, but only found arsenic in one.

Taylor wasn't the only professional gentleman whose Sunday had been disturbed. An attorney who lived in Richmond, Surrey, had been roused on the Sabbath by Dr Thomas Smethurst, whose wife was dying and needed to write a will. He had a draft version, which he said had been prepared by a London barrister. It left everything Isabella had, except for a brooch, to her husband; on her death, he would inherit £1,740 lent on a mortgage. How strange that a married woman, leaving almost everything she possessed to her husband, should feel the need to write a will. But stranger still that it should have been made out in her maiden name, Isabella Bankes, when the couple had been married six months earlier.

Isabella and Thomas had met while living in the same boarding house in Bayswater, London, in 1858. The landlady forced Isabella to leave due to her improper behaviour; Thomas had a wife already. He had married Mary when he was in his early twenties, and Mary was in her mid-forties. By the time Thomas met Isabella, Mary was in her seventies. Isabella was by comparison a mere slip of a girl at 42, the daughter of a sugar refiner who had left her well provided for. She had a life interest in £5,000, which yielded an annual income of about £150 – not an inconsiderable sum, but it wasn't a fortune. When Thomas and Isabella married it was bigamous; Thomas knew, but did Isabella?

Dr Smethurst had gained his licentiate from the Society of Apothecaries in about 1830, just after Taylor had; they were almost the same age. Smethurst's MD was controversial. Some said it came from a European university, but on the 1851 census, he claimed it was from the University of St Andrews. After practising medicine in London, Smethurst had gone abroad, studying hydrotherapy, and he had written a book about it. Back in England, Smethurst ran his water-cure establishment at Moor Park House in Farnham, Surrey. He sold it to Dr Lane, who in 1858 had been embroiled in the scandalous divorce case of Mrs Robinson, one of the first to follow in the wake of the Matrimonial Causes Act. Whether from hydrotherapy or some other source, the Smethursts had enough money to enable them to travel about Europe, before moving into the same Bayswater lodging house as Isabella Bankes.

Thomas and Isabella married in December 1858 and discretely moved to Richmond. Isabella fell ill at the end of March 1859, Smethurst reporting her symptoms as sickness and diarrhoea. She was cared for by the landlady and her daughter, and at the landlady's suggestion, by local physicians Dr Julius and his partner Dr Bird. In mid-April, Isabella's condition deteriorating, they moved to another lodging house, where Thomas ministered to her, refusing to hire a nurse to care for her as he claimed he could not afford one. This flew in the face of evidence that showed he had a healthy bank balance, and Isabella had received her half-yearly income – which Smethurst had in his possession. He invited her sister Louisa to visit, but always stayed in the room, and refused Louisa's offers to sit up with the seriously ill woman. Finally, he turned Louisa away, claiming that Julius and Bird had said her visits were bad for Isabella; this turned out to be a lie.

Isabella asked that another medical professional be sent for, and after some prompting, Smethurst sent for Dr Todd on 28 April. Individually, the three medical men were suspicious of Isabella's symptoms, and each separately came to the conclusion that she was being poisoned. The medicines they prescribed did not work as they should, and they believed that an irritant poison was interfering with them. Her facial expression was peculiar, 'as if she was under some influence or terror', not the result of disease. Dr Todd prescribed opium and copper sulphate, which Smethurst claimed induced a burning sensation from her throat through to the other end of the alimentary canal. The other three doctors said it should not have had such an effect. So, on 29 April, Bird and Julius collected Isabella's stool samples, and passed them to Taylor.

The day after Taylor made his analysis, Smethurst was sent before the magistrates, and was taken into custody for administering poison to his wife. He was given bail, claiming that his absence would kill Isabella; incredibly, the magistrates believed him, seeing no risk to Isabella despite what he had just been arrested for. He went straight back to the lodging house, where a nurse now sat with Isabella, at the direction of the doctors, and Smethurst was 'at liberty' for the rest of the evening and most of the following morning. Had there been arsenic in Smethurst's belongings, he now had the opportunity to discard it.

Isabella died on 3 May, the day after Smethurst was given bail, and he was rearrested. A post-mortem showed blood in her stomach caused by intense irritation, and there was severe damage to Isabella's intestines. It also showed that Isabella was between five and seven weeks pregnant; she had conceived between mid- to late March. Taylor was sent her viscera to analyse, along with many medicine bottles and pill boxes. Perhaps somewhere amongst all these medicines would be the source of the arsenic that Taylor had discovered earlier.

Taylor did not find any arsenic in Isabella's body, but he found antimony: in her intestines, one of her kidneys, and the blood from her heart. Dr William Odling, Guy's professor of practical chemistry, assisted Taylor in analysing the medicines. Ten years earlier, he had been the boy who had tried to deliver the stomach of O'Connor, victim of the Mannings, to Taylor for analysis. Taylor and Odling went through each bottle, testing the contents to make sure they matched up with their labels. They were homeopathic remedies; coupled with Smethurst's career in hydrotherapy, Taylor viewed him as nothing but a quack.

Bottle number 21 was a mystery. It contained a small amount of 'clear watery liquid of saline taste' – Taylor had put some on his tongue. He decided to use the Reinsch test, and so took some out of the bottle, mixing it with hydrochloric acid. He added the copper gauze, and it was 'entirely destroyed, and was dissolved; that seemed rather a remarkable circumstance; at any rate, it had never occurred to me in my experience before.' But just before the copper had dissolved from view, Taylor noticed an indication of arsenic's presence. He dipped in another copper gauze, removing it quickly before it could dissolve, and obtained arsenic crystals from it. Taylor and Odling realised that it was mainly a solution of potassium chlorate, but why did it contain arsenic?

The case went before the magistrates, and then there was an inquest. William Ballantine was hired by Isabella's family, a barrister Taylor had encountered before. The three doctors who had attended Isabella during her illness swore that her symptoms were those of irritant poisoning; the post-mortem appearances indicated the same cause. Taylor agreed; 'I can only account for death by supposing it to be the result of antimony and arsenic administered in small doses at intervals.' Examined both before the magistrates and the coroner, Taylor said that bottle 21 contained arsenic and potassium chlorate. The latter, he claimed, was a harmless saline mixture, which worked on the kidneys; if poison had been administered, it would carry off the 'noxious ingredient' very quickly. Repeated use of it, however, would cause chronic inflammation and would eventually lead to death by exhaustion. He had even included a chapter on potassium chlorate in *On Poisons*. Two possibilities arose: had Smethurst deliberately used potassium chlorate to flush all trace of arsenic from Isabella's body, and had potassium chlorate killed her?

Based on his analysis of the viscera, Taylor said it was more likely that she had died from the effects of antimony, rather than arsenic, even though the theory of slow poisoning by antimony was not universally accepted.

But then Taylor realised he had made a mistake, akin to a police officer discovering a footprint, only to realise that it was their own. It was an error that would haunt him and his reputation forever.

When the Reinsch test was first adopted, its advantage over the Marsh test was the purity of its materials. Zinc, used in the Marsh test, was at risk of being tainted with arsenic. The hydrochloric acid used as part of the Reinsch test was similarly at risk, but was easily tested. The acid and copper would be boiled together, and if a grey metallic film appeared on the copper, fresh acid was needed. Taylor even made his own acid to ensure purity. Copper, however, was guiltless – or so they thought – but it would still be boiled in distilled water before the test, as a precaution. Now, however, it appeared that when copper dissolved in potassium chlorate, any arsenic trapped inside the copper would be freed, and would contaminate the sample. Taylor realised, with a horrible lurch, that the arsenic he had found in bottle 21 was only there *because he had accidentally put it there*.

He tested his copper, first by dissolving some in potassium chlorate, and then dissolving another piece in nitric acid; he found arsenic in both. He realised that the contents of bottle 21 needed to be retested. But there were only a few teaspoons left. Neither Odling, using the Reinsch test, nor Brande, using the Marsh test, could find arsenic in what remained. This confirmed Taylor's fears: that copper was not pure of arsenic, and that by dissolving it, arsenic was set free.

As soon as he realised his error, Taylor reported it. It should not have made any difference to the forthcoming trial, but there were difficulties surrounding the scientific evidence. Taylor had found a small amount of arsenic in one of the motions he had tested, but there was no arsenic found in her body. How could that be? And yet the medical evidence apparently agreed, based on symptoms and appearances, that she had died of irritant poisoning.

Dr Thomas Smethurst stood trial at the Old Bailey on 7 July 1859, for the wilful murder of Isabella Bankes. Sir Frederick Pollock, who had sent Mary May and Mary Ann Geering to the gallows, presided over the trial. Smethurst objected: Pollock and Taylor were friends, 'on terms of intimacy', but his objection was overruled. It had all the signs of becoming a long trial, but as the second day's proceedings were under way, one of the jurymen fell seriously ill. Fortunately, so many of the witnesses were medical professionals that he was soon in the solicitous care of two physicians and a surgeon. The judge discharged the jury, and postponed the trial until the next Session.

A nation held its breath, eager for the outcome of this sensational tale of poison and bigamy. Over a month later, on 15 August, the trial began again, right from the beginning.

Isabella's general health was gone into; some said she was generally healthy, but her sister Louisa said she was prone to bilious attacks, which led to vomiting. The doctors were examined on what they had observed during their attendance on Isabella. When Taylor was examined, he took care to explain what had caused his mistake. Parry, for the defence, pushed him hard, but Taylor's answers were calmer than those he gave Shee during the Palmer trial. Chemistry was in the dock: Taylor's mistake could look like a terrible blunder, but he had to explain that copper's impurity was hitherto unknown, and he had to describe, in terms that the judge and the jury would understand, how the mistake had come about – and crucially, how it didn't affect his other analyses.

It was the potassium chlorate that was responsible, and in twenty-nine years, Taylor claimed, 'I have never met with chlorate of potass in any analysis I ever practised before, and when I do not meet with it the test answers perfectly.' He had used the same kind of copper for twenty years, and had never encountered this problem before. Could acids in the human stomach release arsenic from copper? There was a very limited chance that it could, Taylor replied. But he emphasised that he had made seventy-seven Reinsch analyses during the Smethurst case, and only two of those analyses had yielded arsenic. Taylor would later write that if the risk of arsenic tainting the Reinsch test was as high as was being suggested, then 'it would have led to the constant and inevitable discovery of arsenic in every solid and liquid submitted to examination' – and it clearly hadn't.

Smethurst, like Palmer, furnished himself with an array of scientific witnesses to speak in his defence. Two of them, Dr Richardson and Julian Rodgers, had been witnesses at Palmer's trial as well. They argued that Isabella had died of natural causes, claiming that it was dysentery aggravated or brought on by pregnancy. But the prosecution witnesses, including Isabella's medical attendants during her illness, had more personal experience of dysentery; some had been in India and in Crimea, and they did not believe that was what Isabella had suffered from. Arguing in a similar vein to the Palmer trial, Rodgers said that if someone had died of poison, then the poison should be found. As it hadn't been, then the arsenic Taylor had found in the 'evacuation' had to have come from his impure copper, and Rodgers claimed that pure copper was available. 'I consider it a most dangerous thing to use copper of any other description, particularly where a man's life is at stake.'

A juryman, who later wrote to *The Times*, said that even before Pollock began his summing up, eleven out of the twelve on the jury had made up their minds: Smethurst was guilty. Potassium chlorate and impure copper did not unduly worry them, and the arguments of the prosecution were sufficient to sway them. On the fifth day of the trial, Pollock summed up. Just before the jury retired, Smethurst interjected.

'My lord, will you allow me a few words to clear up some points?'

Pollock refused, saying he must speak via his counsel, but Smethurst spoke anyway. He said that Ballantine knew why Smethurst and Isabella had gone through with their bigamous marriage. As soon as he had been able to, he had written the reason down and sent it to him. And that Isabella's property was no motive at all to kill her. The jury retired, and forty minutes later returned with their verdict: guilty.

Faced with the prospect of a 'scragging' at the hands of Calcraft, prisoners would be allowed to speak. Usually, they would say in a few words that they weren't guilty, or would maintain a stony silence. But not Smethurst, who made a lengthy statement. 'I swear to God that I am perfectly innocent,' and off he went, refuting every piece of evidence made against him. He claimed that he had made great personal sacrifices for Isabella during her illness, he criticised Ballantine again, he pointed the finger at her medical attendants, he claimed Taylor 'says many unaccountable things', he denied Louisa Bankes's sworn testimony. He had told Ballantine that the bigamous marriage was only a temporary measure, for respectability's sake; Mary Smethurst was in her seventies, and as soon as she inevitably died, he and Isabella would be legally wed. Had this really been his plan, they would have had a long time to wait, as Mary lived until 1864.

Pollock responded point for point. Why had he not mentioned before the sacrifices he had made for his wife, or his remarks about the conduct about the doctors who had attended her? How could he claim not to have kept Louisa away when two letters in his handwriting forbade her to visit? And his arguments surrounding the will angered Pollock, that the doctor had 'insult[ed] the ears of the woman who you had apparently made your wife by being called a spinster'. That Smethurst had £70 of Isabella's money in his possession and yet refused additional help disgusted Pollock further, and he declared that Smethurst's statement had merely served to 'add to the evidence given against you'. And with that, Dr Thomas Smethurst, hydropathist and possibly murderer, was sentenced to death.

But Smethurst wasn't finished: 'I declare Dr Julius to be my murderer. I am innocent before God.' And he was led away.

A very lucky escape

There was no appeal court, so a thirty-six-point petition for a pardon was sent to the Secretary of State. Three of the points specifically related to Taylor: his friendship with Pollock, that the impure copper made his positive result in the evacuation unreliable, and that Taylor's deposition before the magistrates had created 'universal prejudice' in the public mind. The petition has a muddled, melodramatic tone, not unlike Smethurst's courtroom statement.

The next day, incredibly, the first Mrs Smethurst lodged a petition to save her erstwhile husband, even though she had been 'grieved beyond all endurance' when he left her for a woman young enough to be her daughter.

It was open season. Everyone had an opinion, and the pages of newspaper and medical journal alike heaved with letters, sent by expert and interested amateurs. A doctor blamed lodging house life, which had left Isabella vulnerable to 'the well-known results of untinned stew pans in daily use'. Mrs Smethurst and Smethurst's brother separately wrote to *The Times*, as did the expert witnesses for the defence. Anonymous clergymen, barristers and doctors aired their views, declaring the evidence to have been too suspect to have warranted a trial.

Seeing an opportunity to capitalise on Taylor's error, and get his own back for the sniping he had suffered from Taylor's pen in the wake of the Palmer trial, Herapath weighed in. He alleged that from the amount of arsenic Taylor had found in bottle 21, he must have dissolved far more copper gauze in it than he would admit to. The tainted copper not only put the evidence of the Smethurst case in doubt, but as Taylor said he had used the same kind of copper for twenty years, 'what shall be said of the justice of the convictions and executions which have taken place during those years upon Dr Taylor's evidence?'

The Home Office were inundated and announced that the case would be reviewed. This fell to Sir Benjamin Brodie, who had been one of the prosecution witnesses at the Palmer trial. He found six points suggesting Smethurst's guilt, and eight his innocence, yet, despite being asked to comment on the scientific and medical aspects of his trial, Brodie included comments on Smethurst's 'indifferent moral character', the issue of inheritance, and Smethurst's financial state. Brodie concluded by saying that he did not think there was 'absolute and complete evidence of Smethurst's guilt', and the Home Secretary wrote to Pollock, saying that he was advising Queen Victoria to grant the prisoner a free pardon. The fault lay, not with 'the constitution or proceedings of our criminal tribunals', but from 'the imperfection of medical science, and the fallibility of judgment, in an obscure malady, even of skilful and experienced medical practitioners'. Pollock replied, objecting, amongst other points, to Brodie wandering outwith the boundaries of his expertise. But the decision had been made: Smethurst was a free man.

For a while, at least; he was convicted of bigamy and served twelve months in prison. On his release, Smethurst was in court again, as Isabella's family challenged her will. He won, and then submitted a petition, seeking compensation for his losses from the murder trial. A note was appended to it: 'It is not very probable that the House of Commons will compensate this gentleman – most people think that he had a very lucky escape.'

Analysts of repute

Taylor did not fare well at the hands of the press. *The Lancet* and the *British Medical Journal* claimed that his speculations were dangerous – that reliance on tiny amounts of poison in analysis was tantamount to the spurious evidence of witchcraft trials. *The Times* chimed in on a similar note. However, Taylor had only been approached in the first place because three doctors witnessed symptoms in a patient that they feared indicated poisoning, and the trial relied heavily on the medical attendants' evidence. And yet, long ago, Taylor had himself criticised Orfila for focussing on tiny traces of poisons that could have arrived in the test sample quite innocently. Perhaps, with the passing of the years, Taylor had fallen into the same trap.

The Lancet were furious that the Home Secretary had blamed medical and scientific evidence for the failure of the Smethurst case, claiming that from the beginning, expert witnesses were hobbled by restrictions put upon their evidence in the courtroom. And that it wasn't science itself that was the problem, but it was one scientist in particular who was to blame: Alfred Swaine Taylor.

Taylor bided his time, gathering his evidence, and in 1860 wrote an article for *Guy's Hospital Reports*. Taylor fully acknowledged that he had made a mistake but, he claimed, it was one that anyone else could have made. He listed 'analysts of repute' who used the Reinsch test, including Herapath without, for once, any viperish remark. But then the tone of his article changed. Although it didn't reach the sardonic heights of the behemoth he wrote following the Palmer trial, Taylor had done his research, and he skewered Herapath, who had criticised not only Taylor's evidence at the Smethurst trial, but had questioned the validity of his entire career.

Taylor had found an article by Herapath in part one of Ure's *Dictionary of Arts and Manufactures*, published in November 1859, where he explained how he performed the Reinsch test. Herapath did not mention testing the purity of the copper, which Taylor used to illustrate that even Herapath wasn't aware that it could be tainted with arsenic. Herapath told readers that he used No. 13-sized copper wire, so Taylor procured some from 'three of the first London chemical dealers'. He tested it and cheerfully announced that each sample contained more arsenic than the copper gauze he and Odling had used in the Smethurst case.

Several experts had come forward to say 'that there was no novelty in the announcement of this universal presence of arsenic in the so-called purest forms of copper'. Whilst it was known that arsenic was associated with copper ores, it was thought that arsenic was 'evolved' out of copper during the smelting process because it was more volatile. Taylor listed a huge number of recent 'works of great chemical authority', and only one, the German-to-French translation of which was published in 1858, mentioned that copper might be tainted with arsenic. It was not a fact that was easily accessible to British scientists so, Taylor claimed, it wasn't a universally known fact at all. But what was well known, and Taylor chose not to dwell on this, was that copper dissolves in potassium chlorate, as well as other substances, and that in those situations, a different test than Reinsch's should be used.

Continuing his copper-testing spree, Taylor acquired samples of copper 'from five gentlemen, of great repute in the three kingdoms, as analysts in cases of poisoning'; we might assume this included Christison and Geoghehan. He found arsenic in all of the samples, and in two cases, in greater proportion than in the copper he used himself. He roped in chemist Dugald Campbell, and the twenty different types of copper Campbell tested also contained arsenic. Rodgers and the other defence witnesses at the Smethurst trial had claimed that arsenic-free copper was easily available, in order to undermine Taylor; but he had shown that they were wrong.

Taylor, who had feared that without the threat of toxicological detection, poisoners would rampage through the land, was now faced with the terrifying prospect that the public would no longer trust him or his tests. That the metallic deposits on his copper, and the white crystals in his test tubes, would be considered by juries to be nothing but the results of blunder. More seriously, it would give ammunition to argumentative counsel to shoot down toxicological evidence. So Taylor described methods for testing the purity of copper, and explained that another type of test should be used when dealing with chlorates to prevent the arsenic locked inside the copper from escaping. Rodgers's claim at the trial that Taylor's positive result for arsenic in the stools was from his tainted copper was probably wrong – and Rodgers likely knew this.

For all that Taylor had got himself – and toxicology – in a dreadful muddle, it must be borne in mind that a guilty verdict at a poisoning trial often resulted in a combination of factors: symptoms, motive, opportunity, post-mortem appearances, as well as toxicological results. Even though Ballantine commented in his memoirs that the Smethurst trial showed why 'the speculations of scientific men, however eminent, ought never to be made the basis of a case', a guilty verdict was never solely the result of a dull grey sheen on a piece of copper.

Letheby was pushing for a change in the way the legal system dealt with scientific evidence, and he teamed up with sanitary chemist Robert Angus Smith. In January 1860, their campaign led to an invitation to speak before the Royal Society of Arts. Smith said that in court, a scientist simply becomes 'a barrister who knows science' and that even with the most careful pre-trial preparation, scientific truth is confounded by examination in court. Smith urged that the scientific expert should be an independent voice, 'not a mere tool of a barrister or his client'. The experts discussed the issue, some saying that it was very wrong how a number of scientists coined in vast incomes from distorting their opinions to suit their clients; there was much money to be made in patent cases and representing companies in environmental disputes. Equally, the legal profession came in for criticism for distorting scientific evidence.

Taylor spoke, agreeing with Smith's idea of having independent scientific assessors. He lamented that he often came to court not having heard the full picture. Considering that Letheby had aided in bringing this discussion about, it says much that Taylor was willing to contribute his opinion. Perhaps both men wanted to iron out the problems they had faced, and to do so in a far more constructive manner than they had before. They had managed to sit in the same room without hurling sarcastic retorts at each other, which was a start.

Not that Letheby and Smith managed to push through any changes at once. But the grievances of the scientists had had a proper airing, and change would eventually come.

Silver with the hue of crime

Taylor had fearsome critics, but he also had his fans. Philosophising on the subject of small talk in Dickens's magazine *All the Year Round* in 1860, Charles Allston Collins advised on desisting from talking about other people's professions: 'I would as soon think of engaging with a gamekeeper, or a horse-dealer, in half an hour's chat on shooting or horse-keeping, as I would of discussing medical jurisprudence with Professor Taylor.' As far as Collins – and perhaps most of his readers – was concerned, Taylor was still the expert.

Two years later in the same magazine, Dickens wrote a somewhat psychedelic interview with Taylor, where the toxicologist appeared in character as twelfth-century alchemist Artephius. Dickens, who was fascinated by the detection of crime, had visited Taylor in his laboratory at Guy's. He had been shown various tests, as well as several body parts, such as dried flakes of human liver in a jar, and a stomach in a fume chamber.

It should have been interesting enough, but Dickens dressed it up in purple prose: the Marsh test involves 'the fairy hydrogen', who forces out arsenic, turning the apparatus 'silver with the hue of crime'. The test for strychnine is assisted with 'imps' – this section suggesting that Taylor had improved his methods for detecting that particular poison. Reading Taylor's deliberately clear scientific explanations, garbed in Dickens's overwrought language, makes for an unusual experience, yet his typical arguments persist: if an analysis is made quickly enough, 'there is no cunning that shall master ours.'

Wilkie Collins may have already crossed Taylor's path during the Palmer trial; perhaps Anne Palmer's 'thin, white dress' inspired Anne Catherick's garb in *The Woman in White*, and maybe Mrs Palmer even lent her first name to the character. Certainly the Palmer cases had inspired Charles Felix, author of the first English detective novel, *The Notting Hill Mystery*, published in 1862. Concerning a wife poisoned with antimony, whose husband has insured her life with several assurance companies, the parallels are all too obvious.

Collins's overlooked Sensation novel *Armadale* was serialised from 1864 to 1866. Its anti-heroine, Lydia Gwilt, is in her mid-thirties, and passes herself off as somewhat younger in order to marry a man in his early twenties – perhaps inspired by Thomas Smethurst's first marriage.

Her background is murky, having allegedly murdered her first husband with poison, and after being convicted, is granted a pardon by the Home Secretary. Collins implicitly invokes the Smethurst case when he explains how Gwilt 'was tried all over again, before an amateur court of justice, in the columns of the newspapers. All the people who had no personal experience whatever on the subject seized their pens, and rushed (by kind permission of the editor) into print. Doctors who had not attended the sick man, and who had not been present at the examination of the body, declared by dozens that he had died a natural death. Barristers without business, who had *not* heard the evidence, attacked the jury who had heard it, and judged the judge, who had sat on the bench before some of them were born.'

The novel ends with Gwilt attempting a murder using carbonic acid – even if it is scientifically inaccurate and she only manages to poison herself. During *Armadale*'s run, three sailors died aboard a ship moored at Liverpool, from 'sleeping in poisoned air' – and the ship's name was *Armadale*; the sort of astonishing coincidence that stretches credulity even in the pages of Sensation fiction.

Collins noted that 'Wherever the story touches on questions connected with Law, Medicine, or Chemistry, it has been submitted, before publication, to the experience of professional men.' A friend had sent him a plan of the apparatus that he had Gwilt use for her carbonic acid murder; he even 'saw the chemical ingredients at work'. Given that Dickens and Collins were friends, there is every chance that this could have been Taylor. Collins doesn't name the professionals he sought advice from, but even if he didn't speak to Taylor in person, Collins owned the first two editions of *On Poisons*. Poisons crop up constantly in Collins's work, and he wouldn't be the last author who would look to Taylor for inspiration.

Chapter 15

You are the Villain
1860–62

That poor boy

When Taylor was called in to help investigate a suspected crime, he was a distant observer of a community with which he had no association, the victim unknown to him, their family grieving or guilty strangers. But in 1860, a brutal death that scandalised Victorian Britain befell one of Taylor's own relatives.

Reginald Channell Cancellor was the son of John Henry Cancellor, eldest brother of Taylor's wife. Cancellor was a barrister, and Reginald's middle name came from his godfather, the judge William Fry Channell. There was a gap of eight years between Reginald and his next eldest sibling; he was presumably a surprise when he arrived in August 1844 at Chester Terrace, Regent's Park, just around the corner from the Taylors' home. Edith, the Taylors' daughter, had been born a couple of months earlier, and the cousins may have been brought up closely when they were young.

As a medical professional, Taylor must have realised that Reginald had a disability. His head was unusually large, 'peculiarly formed', and he was short for his age and rather stout. He was stubborn and lazy, and seemed impossible to educate. Reginald's own brother, Reverend John Henry Cancellor, said, 'He would not readily do what he was told. Figures were a weak point of his.' Reginald's parents were determined that the boy should be educated; he was sent to a private school in St Leonard's, Sussex, then had a private tutor. Even if Taylor had intervened and suggested that Reginald might be incapable of much learning, it seems that no one would have listened.

Reginald's reverend brother received a recommendation for Thomas Hopley's school in Eastbourne, from no less a person than the chief constable of the Sussex Constabulary. Hopley had grand ideas on educational reform, and a reputation for dealing with difficult boys. He had spent his early years studying the conditions of child and female labour in factories, and had investigated ragged schools, prisons and asylums, cranking out pamphlets to trumpet his ideas for reform.

In desperation, at the end of 1859, the Cancellors packed Reginald off to Hopley's school. When he came home for Christmas, the family believed that Reginald had improved, but his father complained to Hopley that too much violence had been used against him. Hopley charged £180 a year in fees, and presumably Cancellor thought the headmaster would listen to him. Despite Reginald's stubbornness, which the family was so eager to overcome, he 'was a particularly kind and affectionate boy, and very much attached to his father'.

In April, Hopley wrote to Cancellor. The boy was becoming obstinate again. Hopley didn't like to use physical punishment against his pupils – he didn't have a cane, that traditional instrument of barbaric school discipline – but he sought Cancellor's permission to use 'extreme punishment' in order to force a permanent change in Reginald's attitude. Cancellor, who had no doubt been beaten himself as a boy, approved. He had no idea that he had just signed his son's death warrant.

On the evening of 21 April 1860, 15-year-old Reginald had been in the playroom with the other pupils, when Hopley called him into the dining room to do a maths lesson. But the obstinate boy refused. Hopley, sanctioned under British law (which was not entirely scrapped until 2003), exercised corporal punishment on him. Without a cane, he beat the boy with what came to hand, which happened to be a skipping rope and a walking stick. Reginald screamed, 'loud enough to be heard half over Eastbourne', Hopley later confessed. He paused to give Reginald a chance to say his lesson, Reginald refused, and he beat him. This pattern continued for two hours. Not long before midnight, Hopley took Reginald up to his bedroom. Reginald could barely walk, and Hopley had to help him up the stairs. The beating continued.

It finally stopped just after midnight. Two servants later reported the noise of much coming and going by Hopley and his wife. At 6.30 am, Hopley returned to Reginald's bedroom and, so he said, discovered that the boy was dead. Hopley sent a telegram to Cancellor, who came at once and viewed his son's body, which had been washed and carefully dressed, so that only his face showed. The local surgeon based his diagnosis on information from Hopley, saying that Reginald had died of heart disease. Cancellor asked him, perhaps with guilt, if 'any slight punishment' could have had an effect, and the surgeon thought it wouldn't have.

An inquest was held, and the surgeon's opinion went unchallenged. But rumour was afoot, thanks to the servants, who had heard Reginald's screams, and they said they had seen blood on floors and furnishings, on Reginald's clothing, as well as that of Hopley and his wife. The Hopleys' nursemaid, who had heard the beating from the bedroom beside Reginald's, wrote to the sister-in-law of Sir Charles Locock, a physician who attended Queen Victoria. He was an acquaintance of the Cancellors, and probably knew Taylor as well. Locock contacted Hopley about Reginald's death, and the headmaster paid him a visit, asking if he would help to clear up the rumours if he issued a certificate saying that Reginald had died of heart disease. Hopley confessed to the beating, but when Locock asked about the blood, Hopley 'turned pale, and made no reply, and, after standing still for a minute or so, he took up his hat and went out of the room.'

Reverend Cancellor went down to Eastbourne with an undertaker, to collect his brother's body. He told Hopley that there would be a post-mortem, and he quizzed him. Not for nothing was he the nephew of a leading forensic scientist. Had he drawn blood? Had there been any noise after 10.00 pm? Why had he told their father that Reginald had been happy after the punishment? It was clear that Hopley had lied to their father, and Hopley told the young priest that he had only lied in order to spare him. He showed him Reginald's body lying in a lead coffin, and Reverend Cancellor asked Hopley if the mark on Reginald's face was a bruise. Hopley denied it, saying it was normal discolouration after death. Looking down at the body, Hopley 'clasped his hands together and said, "Heaven knows, I have done my duty by that poor boy."'

The post-mortem was carried out by a surgeon and two doctors at St George's Hospital in London. It could not be done at Guy's – Taylor could not be involved. Reginald was wearing kid gloves, and long stockings that came up to his thighs. When the body was uncovered, the skin was 'a dark livid colour', and the post-mortem showed that there was a huge amount of blood under the surface. There were two wounds on his legs, which could have been made by the pointed end of the walking stick. The flesh of the thighs had been reduced to jelly, so it was no wonder that Reginald had struggled to walk up the stairs. They remarked that the shape of Reginald's head suggested that he had had hydrocephalus, and this was confirmed on opening it up. Learning difficulties can be a consequence of congenital hydrocephalus; perhaps Reginald had not been an obstinate boy at all, and had suffered from a disease that at the time was untreatable and misunderstood.

A second post-mortem was carried out by John Erichsen, a surgeon at University College Hospital. In his view, it seemed that Reginald might have had a blood condition that wasn't unlike haemophilia. This would put a very different slant on events; if Hopley had used reasonable force, Reginald could have died from complications caused by an underlying medical condition.

Ballantine was hired for the defence by Hopley's brother, an eccentric artist of the barrister's acquaintance who had to sell his painting *The Birth of the Pyramid* to pay his brother's fees. Ballantine took the brief with great reluctance as he was a friend of Cancellor's; they were both serjeants-at-law, a now defunct order of barristers much like King's or Queen's Counsel. Ballantine believed that Hopley was guilty, as made abundantly clear by the way he later wrote about it: 'Screams that sent terror into the hearts of the other pupils were heard during the long hours of the night. The poor children lay shuddering with fear and horror, whilst the wretched half-witted victim of a lunatic's system of education was deliberately mangled to death.' To Ballantine, it was murder, and he wanted to claim that

Hopley was not guilty by reason of insanity. Instead, Hopley would be tried for manslaughter; he had done nothing illegal in beating a schoolboy in his charge, but it was up to the jury to decide if he had exceeded reasonable force.

John Henry Cancellor died in June, a month before the trial. It isn't clear if the stress he and his family were under, not to mention the guilt he must inevitably have felt, had killed him, or if he was already unwell, and the terrible circumstances of his son's death exacerbated his illness. In the space of two months, Taylor had lost a nephew and a brother-in-law. The great strain may have taken its toll on Reginald's mother as well, who died only two years later.

Hopley was tried at Lewes Summer Assizes. The judge was Lord Chief Justice Cockburn, who had led the prosecution at the Palmer trial. Although Ballantine had apparently prepared some of Hopley's pupils to speak in praise of their teacher, none seem to have given evidence, and he didn't call Erichsen. The pupil in the room next to Reginald's slept soundly all night, suggesting that Reginald had not screamed. This might indicate that Hopley had not beaten him very hard, but Ballantine did not call him either. Although he wrote in his memoirs 'I hope I did my duty to him as his counsel,' it seems as if Ballantine's own opinion of Hopley's guilt, coupled with his acquaintance with Reginald's now dead father, affected his handling of the case.

The prosecution called on the servants, and even the evidence of three coastguards, who said they saw lights on in the house late that night. What could this have proved, unless they were furnished with telescopes and had seen Hopley in the act through a window? Cockburn declared that 'By the law of England, a parent … may for the purpose of correcting what is evil in the child, inflict moderate and reasonable corporal punishment.' The jury decided that Hopley had exceeded it, and found him guilty of manslaughter. Cockburn sentenced him to four years of penal servitude.

Public opinion was very much against Hopley, and the press portrayed him as a monster. *The Times* said that 'to beat a boy for two hours with a thick stick and a skipping rope, to macerate him, to "prod" him, in private and at midnight, is not discipline, but murder.' A poem, imaginatively titled *Beaten to Death*, appeared, which opened:

> *At depth of night, this thought on home had shone;*
> *'Our distant child draws safe his sleeping breath.'*
> *E'en then the cherish'd boy, th' expected son,*
> *Was dying through two hours – beaten to death.*

On it goes melodramatically for ten more mawkish stanzas. But 'The Eastbourne Manslaughter', as it was known, revealed very Victorian anxieties. The middle and upper classes, who sent their children away from home to be educated, had to face the fear that their children might not be safe. And Cockburn's declaration was used as a test into the twentieth century, even being quoted as recently as 2004 in a House of Lords debate on reasonable chastisement of children.

Taylor's *Manual of Medical Jurisprudence* went into its seventh edition in 1861. He could not omit the famous Hopley case, even though it was personal. It does not seem that Taylor's name was linked with the victim's in the press, even though it was made no secret that the boy's father had been an important barrister, and his godfather a judge. Taylor tried to write dispassionately about it, but the case crops up three times in the course of the book. He includes his nephew's death in the section on 'Wounds – Death from Flagellation':

[Hopley's] guilt was established by the fact that he had endeavoured to conceal the effects of his violence by removing the marks of blood – that he had covered the body of the deceased with clothing so as to conceal the bruises, – that he had procured a coroner's inquest to be held in haste, and while concealing from the jury the fact that he had beaten the youth on the night of his death, stated that he had found him dead, and suggested that he might have died of disease of the heart.

The bereaved and angry uncle is visible beside the sober scientist.

Spare no labour

When the census was taken on the night of 7 April 1861, none of the Taylors were at home. Caroline and Edith were staying in Hendon with one of Caroline's brothers. Taylor was staying at the Queen's Hotel in Chester, where the Lent Assizes were in progress. Amongst the guests were Rupert Potter, a barrister who would become the father of children's author Beatrix Potter.

Taylor wasn't called to give evidence at the Assizes, though there were several cases that he could have given an opinion on. There was manslaughter from a broken neck, cutting and wounding, and a mother accused of manslaughter for not breastfeeding her child; she was acquitted on medical evidence.

At the end of the 1850s, after the Palmer and Smethurst trials, it may have seemed as if Taylor's reputation was in tatters, his career at an end. But he wasn't sacked from Guy's Hospital, and his qualifications weren't revoked from the Society of Apothecaries, or the Royal Colleges of Surgeons or Physicians. The University of St Andrews did not rescind his honorary degree, and John Churchill still published Taylor's books. He was extremely busy in the 1860s, more active than ever at an age when most men of means would have packed away their Bunsen burners and retired.

In the year that his nephew was killed, Taylor examined blood in Penrith, and suspected poisonings in Norfolk, Essex, and Liverpool. He was able to prove that the death in Essex was from natural causes, but as the death was rumoured to have been caused by poisonous herbs administered by a quack doctor, the coroner used it as an opportunity to warn the man of the dangers that his 'medicines' could cause. The case in Liverpool involved antimonial poisoning again. It was proving difficult to convince juries of, as Taylor saw it, antimony's dangers and the defendant was acquitted after a three-day trial.

At the end of the year, Edward Cresy wrote to Charles Darwin about the sensitivity of tests for arsenic, and referred to Taylor's *On Poisons*. Taylor wrote to Cresy, pointing out to both men a piece in *Chemical News* about detecting saline 'in an almost infinitesimal state of division'. The letter is careful and scientific, except for its ending. When referring to his article on arsenic impurities in copper, Taylor remarks like a schoolboy, 'I am glad you think that Herapath has undergone a proper rasping in my paper.'

When Taylor wrote the preface to the seventh edition of his *Manual of Medical Jurisprudence*, he couldn't help but mention that, since the first edition back in 1844, 15,750 copies had been printed. 'This unexpected encouragement has induced me to spare no labour in order to maintain the work on a level with the progress of medical and legal knowledge.' He decided to keep in additions made to it by the Scottish judge, John Hope, who had died in 1858; as Scotland has a different legal system to England and Wales, it meant that Taylor's book was wider in scope. Along with its large section on poisons, it covered the typical range of forensic medicine concerns: wounds, poison, sexual assaults, death by cold or lightning strike, all compiled from sources from around the world. There are many examples of Taylor having been contacted personally for his opinion on cases, in Britain and overseas, such as a sodomy trial in Jamaica in the late 1840s. No wonder he was so concerned when he discovered that employees of the Royal Mail had been nosing through his letters.

Witchcraft at Wimbledon

1862 commenced with a headline that appears to be straight from the pages of Sir Arthur Conan Doyle: 'Professor Taylor and the Case of Witchcraft at Wimbledon'. In February, Selina Smith, a 20-year-old gypsy, had arrived at the Kings' middle-class residence in Wimbledon, to sell mats. She told the servants that she read fortunes. For sixpence, she told the cook that she would marry a gentleman, and that she 'had a lucky countenance'. Later, the magistrate, Mr Dayman, asked with heavy sarcasm, 'I suppose you thought that was worth sixpence?' Selina gave exactly the same fortune to one of the other servants.

As Mrs King had told the cook she should like to have her fortune read, Selina was sent up to the mistress. Cook introduced her as a white witch, and Mrs King paid Selina with money and an old dress. Regarding Mrs King's husband, the cook heard Selina prophesy, 'As sure as God, he will die in a month.' Mrs King was heard to say that she would pay a sovereign for something to do her good, and Selina said she would bring her something. Three drops in Mr King's tea would mean his death in a month.

Selina reappeared at the house with some brown powder. She had the cook mix it in a bottle with cold water, and then asked Mrs King ten shillings for it. Mrs King wouldn't pay her, and Selina refused to leave. Mr King, who heard a fracas going on from the gentlemanly comfort of his study, left the house, and returned with Sergeant Davis of V Division. Davis arrested Selina on the spot for taking two shillings and a dress by false pretences.

Selina was examined before Wandsworth police court. The cook was laughed at when she reported Selina's fortune, and when Mr King came forward, he said that he was Mrs King's husband, 'unfortunately', and called it all 'silly business'. The fact that his wife had apparently been willing to poison him was easily dismissed.

Sergeant Davis took the suspect bottle to Guy's Hospital, but Taylor refused to be involved 'until he was assured that his expenses would be paid'. He had made the same protest thirteen years earlier during the Manning case. Taylor said that he wanted five guineas for the analysis, and another five for his appearance before the court, 'besides other expenses he might be put to'. He complained that his fees in other cases hadn't been paid, and he 'thought it was time that they were properly classified'.

Indeed so – but it left a problem at the centre of the case. If they wanted Taylor's analysis, they would have to apply to Sir Richard Mayne, who would submit the request to the Secretary of State. And Dayman didn't think the bottle contained anything of a 'fatal character', which was easy for him to say, as his wife hadn't been about to sprinkle it in his tea. Selina offered to prove it was harmless by swallowing the entire contents of the bottle herself, but Dayman wouldn't allow it, 'for in case she were poisoned, they would be accessories to her committing suicide'. So Dayman and King asked their 'chemical friends' to test it.

'Medicus' wrote to the *Morning Chronicle* to tell the world an unlikely story, that gypsies had a secret, untraceable poison called drei or dri, a fungus that was inactive until swallowed. It would then cast out 'innumerable greenish yellow fibres about 12 or 18 inches in length', and the victim would suffer fever, a cough, spitting blood until death claimed them two to three weeks later. Two days after death, no post-mortem would show any trace of the fungus. It was unsubstantiated and irresponsible, only increasing public distrust of gypsies.

Mr King wanted a swift end to the 'silly business'; he did not want to be exposed 'to any further annoyance', having received several unpleasant, anonymous letters. He had taken the dubious step of examining the bottle himself, and announced that there was nothing harmful in it, and 'without mentioning names, he was convinced that no evil intentions were entertained.' Quite what went on behind doors between Mr and Mrs King afterwards is unknown.

Selina was punished for fortune telling, and sentenced to three months' imprisonment with hard labour. If only all the cases Taylor was involved in were so easy to joke about: the next was one of the most sinister of his career.

The greatest criminal that ever lived

The trouble with being a serial killer is that eventually you'll be caught. Constance Wilson, alias Catherine Wilson, alias Catherine Taylor, had ingratiated herself into many homes, and inveigled many sums of money from unsuspecting, trusting people. One such person was Sarah Carnell, who fell ill during a visit from Constance in February 1862. Sarah ran a lodging house in Marylebone, and had known Constance for several years by the name of Mrs Taylor. On at least one occasion, Sarah had fallen seriously ill with vomiting after eating at Constance's house. This time, Constance said she would fetch Sarah some medicine. Sarah was vulnerable at the time, in the middle of separating from her husband. Although she had specifically asked for a rhubarb mixture, Constance had returned with a 'black draught' (magnesium sulphate), saying it was soothing, and that the doctor wouldn't give her rhubarb.

Sarah held a glass while Constance poured the liquid in, and Sarah found that it was very hot. 'Drink it up, love,' Constance said. 'It will warm you.' Sarah put it to her lips, but didn't swallow. It badly burnt her mouth, and she spat it out on the bedclothes, leaving them full of holes. Sarah's son came in, and she told him that she must have been given the wrong medicine. Constance insisted on staying with Sarah, looking after her as she continued to feel unwell, and slept in Sarah's room.

Constance left the next day. A letter arrived, saying that she had been run over by a cab; it wasn't signed by Constance, but it seemed to be in her handwriting. When Sarah went to Constance's house, she found it boarded up, and all the furniture was gone. Sarah went to the police, and D Division's surgeon saw the effects of oil of vitriol (sulphuric acid) in her mouth. When Constance was tracked down by the police, she was walking arm in arm through Camden Town with Sarah's husband.

Constance was tried at the Old Bailey in June 1862, charged with attempting to poison Sarah Carnell. She was defended by Montagu Williams, who found

Constance's appearance 'a very peculiar one, her chin being the most receding one I have ever seen'. The defence argued that the chemist had dispensed the wrong medicine, but the chemist's assistant said that he had definitely given Constance black draught, and that there was nothing in it which would 'burn a person's mouth or sheet'. He said that if you mixed sulphuric acid with anything, even water, it would immediately become hot. The problem for the prosecution was that Sarah thought the cup she had been given by Constance was empty, but she admitted the light was poor in the room, so she couldn't be sure. The black draught couldn't have been already mixed with the acid in the bottle, because it would have been too hot for Constance to hold.

While the jury deliberated, a stranger placed his hand on Williams's gown, and said, 'Very ingenious, sir, but if you succeed in getting that woman off, you will do her the worst turn that anyone ever did her.'

Williams was astonished, and discovered that he was speaking to an officer from Lincolnshire County Constabulary. If Constance was acquitted, the officer had instructions to arrest her on seven separate charges of wilful murder.

The jury came back with a verdict of not guilty, and 'an expression of delight' passed across Constance's peculiar face. She had no idea that her past crimes had caught up with her, and just as she left the dock, she was arrested.

The coverage of the Sarah Carnell case had impelled people to come forward. There was the death of Captain Mawer in Lincolnshire; Constance had been his housekeeper. There was James Dixon, 'a young gentleman', whom she had passed off as her husband. Two months after Dixon's death, a Mrs Soames had died – she had run a lodging house where Constance had lived. And then there was Ann Atkinson, a wealthy woman who Constance had ingratiated herself with. These cases barely scratched the surface. Thomas Nunneley from Leeds, who had been one of the expert witnesses selected by the defence at the Palmer trial, was tasked with performing the chemical analysis in the Atkinson investigation. Taylor, faced with a jar of remains 'sodden and white and converted to adipocere', was analysing Mrs Soames's viscera, as well as Mawer's.

As ever, Taylor's skills were in demand. He'd been approached by the Cornish authorities for bloodstain analysis in the investigation of the death of a wealthy grocer. In August, during the Wilson investigations, Taylor travelled to the Summer Assizes in Bodmin, 260 miles from London. He had found blood on clothing, and a billhook, and a bloody footprint on a floorboard. During the trial, the defence counsel asked Taylor's opinion on rigor mortis and time of death. Taylor's evidence, added to that of other witnesses, led to the execution of John Doige. It was hangman William Calcraft who once again flung a miscreant into eternity. Would he be preparing his rope for another murderer?

A distinct *modus operandi* emerges when the cases involving Constance aka Catherine are laid out together. She would ingratiate herself, make claims to poverty, borrow money or accept it as a gift, and would keep extracting money until her victim had to borrow money themselves. She would fake letters – either from herself, describing entirely fictitious situations, or claim that the letters were from real people, or from people who did not exist at all. When her victim fell ill – perhaps the illness was genuine, or brought on by a whisker of poison – she would take over the nursing. She had been trusted by Mrs Atkinson, and she told Mrs Atkinson's husband and nephew that the woman had fallen ill after being attacked and robbed in the toilets at Rugby station on the way to London. She had been trusted by Mrs Soames, who was always in good health, but had suddenly fallen ill, Catherine keeping Mrs Soames's curious daughters and lodgers away during her suspiciously short illness. But under the rules of criminal pleading, the jury were not to be told about the other cases; the defendant could only be tried on one case at a time.

She was tried under the name of Catherine Wilson at the end of September 1862 for the death of Mrs Soames six years earlier. The trial lasted for three days. The witnesses were Mrs Soames's relatives and lodgers, who remembered Catherine moving into Mrs Soames's house, and recalled the circumstances of the dead woman's illness. It became clear that not only had Catherine forced Mrs Soames into lending her so much money that she had had to pawn her milk jug and borrow money from her half-brother, but that Catherine had been telling lies. She had spread a rumour that Mrs Soames had experienced a romantic disappointment, and committed suicide as a result.

There was important evidence from a lodger called Emma Rowe, who had been very close to Mrs Soames. She 'used to tell me everything, up to within six weeks of her death nearly' – and then Emma had been supplanted by Catherine. Although there had been an inquest into Mrs Soames's sudden demise in 1856, her death was attributed to natural causes. As far as her medical attendant was concerned, the symptoms indicated English cholera. But Emma Rowe had not been called to give evidence. Had she been, then Catherine's criminal career might have been cut short.

Emma had seen two medicine bottles in Catherine's possession, one of which she called 'Mrs Soames's medicine'. Catherine had said 'it was particular stuff, and it must be given by herself'; she would not allow anyone else to administer it. When Mrs Soames died, Catherine spun her lie, saying that the dead woman had been deceived by a man who had borrowed £80 from her and killed herself because 'she was tired of her life'. Emma challenged her. Why hadn't Catherine said anything to the family, or the medical man? Catherine replied that it wasn't her business, and Emma said, 'You are the villain.' Catherine did not reply.

Not only had a crucial witness been left unexamined at the inquest, but the original medical investigation had been botched. Whidborne, the surgeon, had heard rumours that Mrs Soames had been poisoned. Rather than send the whole of the stomach for analysis, he had sent only the stomach contents. The viscera were eventually examined, but Whidborne said, 'I did not make any notes of the examination – my evidence before the coroner was given from recollection.' How Taylor must have rolled his eyes.

And now, six years later, Taylor was performing the analysis. Mrs Soames's insides had turned to an adipocerous mass of soap, so he couldn't see anything in the appearance of the now waxy organs to indicate poisoning. He had needed extra time to perform his analyses, presumably because, if it was impossible to tell the organs apart, he would have had to chop up the entire remains into small pieces to be able to run his tests on them. He found no poison, but this didn't surprise him. The fact that the body had turned to adipocere suggests that it had been in a damp grave, and this could very well have washed out any poison, had it been present.

It was thought that Catherine had used colchicum, as there was evidence from another case that she had talked of using it for rheumatism. It was a medicine derived from a flower, but could be poisonous. At the trial, Taylor explained that Mrs Soames's reported symptoms could have been those of colchicum poisoning: 'a sense of fullness and heat in the throat, and violent pain in the stomach' had been described in some cases. As Taylor usually claimed, when faced with a difficult case of organic poisoning, it was impossible to find any organic poisons after so long.

During the cross-examination, a personal tragedy was revealed. On being asked how quickly the symptoms of Asiatic cholera came on, Taylor replied, 'In the case of a friend of mine, who died from it, they came on in seven or eight hours; he was dead in about nine hours, from a perfect state of health.' Taylor had to admit that he hadn't attended many cholera patients, and hadn't practised medicine for some years. But he made up for this by declaring to the court that 'for many years I have confined my professional practice to the subject of poisoning, to jurisprudential matters.'

Taylor emphasised the fact that English cholera could easily be confused with poisoning. He had 'in my mind at least six cases which have been registered as death from cholera, which have been from poison'. Thinking some more, Taylor increased this number to eight cases. When there was no epidemic, and death is registered as cholera, 'I should immediately suspect poisoning.' A startling claim, but Taylor said he would require an analysis to be performed 'before I gave an opinion'. As Whidborne had prescribed a medicine that usually gave patients of English cholera relief, and as it hadn't worked, Taylor said this further suggested that Mrs Soames hadn't had English cholera at all.

Thomas Nunneley, who was ready to give evidence at the trial for Mrs Atkinson's murder if Wilson was acquitted for Mrs Soames's murder, mostly backed up what Taylor had said. But, the echoes of the Palmer trial lingering still, when asked if he agreed with Taylor, he said, 'Generally – the answer with which I should most disagree, would be as to the impossibility of finding any vegetable poison.' But he entirely agreed that finding colchicum in particular after so long would be impossible.

Wilson was found guilty. After the trial, the judge sent for her defence counsel, Montagu Williams, and congratulated him on his performance, but said that the facts were too strong. He recounted the criminals he had prosecuted and defended in his time as a barrister, and then said, 'In my opinion you have today defended the greatest criminal that ever lived.'

The Times was astonished by what Taylor had said about English cholera. 'When we are assured by Dr Taylor that numerous cases of death, attributed to cholera, are in fact occasioned by poison, it is high time that the attention of the medical and legal professions should be directed to this subject.' But this was eight cases over a long career; it would include Sarah Chesham's sons, Thomas Jennings's children, Ann Bacon and Ann Palmer. Taylor was not necessarily encouraging panic, especially as he said that he would need to perform an analysis before forming an opinion. But barrister W.S. Austin, writing in George Augustus Sala's magazine, *Temple Bar*, used it as a weapon against him.

In writing about the Wilson trial in his series of articles on circumstantial evidence, Austin, in an article called 'Secret Poisoning', said that Taylor's comment was 'evidently meant to put the medical profession and the public on their guard', and it was Austin who put the phrase 'secret poisoning' into Taylor's mouth. Although it was an expression that Taylor had not been economical with in the past, he does not appear to have used it at the Wilson trial. Austin said that Taylor was implying that many cholera cases 'are in reality cases of secret poisoning'. It seems to be Austin who was exaggerating, perhaps because other parts of the press, such as *The Times*, had already done so in their coverage of the sensational trial. Austin said that it was an 'astounding assertion', and that Taylor was so obsessed with secret poisoning that it 'blinds his judgment and weakens his logical power. In one sense, much erudition makes him mad. He rides his own hobby to death.'

Austin said that he knew Taylor's books as 'most advocates know them – as valuable works of reference – and I have the highest opinion of his intellectual acuteness and of his extensive and varied reading,' but Austin said that Taylor 'has never struck me as possessing the logical faculty in a high degree'. In a move perhaps calculated to enrage Taylor, who hated all quackery, Austin described Taylor in phrenological terms: 'while his eye has the brightness and restlessness of a man of genius', bumps on his head showed that he lacked reflection and reasoning. Austin was critical too of Oppenheim, who had defended Wilson with Montagu Williams. How much they took Austin's criticisms to heart is unknown: they might have disregarded it. At the time, *Temple Bar* carried the serialisation of Mary Elizabeth Braddon's Sensation novel *Aurora Floyd*, and Sala's own fiction, under the ludicrous title *The Strange Adventures of Captain Dangerous*.

Catherine Wilson, or Constance Wilson, or Catherine Taylor, met her end on 20 October 1862 outside the Old Bailey, before a crowd 20,000 to 30,000 strong. She had tried to gain a reprieve, claiming that there was no proof that Mrs Soames had died of poison, or that Catherine had bought any, and that none had been found in her possession. It was rejected. No one made any attempt to plead for her life, no Quakers arranged a petition. 'All consideration of sex, which in cases of women under capital sentence induce some people to plead for mercy, appear in this instance to have been completely merged in horror at the crime, and not a finger was raised to deprecate a just retribution.' She was hanged by William Calcraft, one of the few people in Britain who had killed more people than she had.

Chapter 16

Blood Enough

1863–70

Masters of the science

In early 1863, Taylor and his friend Professor Brande wrote a textbook called *Chemistry*. The *British Medical Journal* had nothing but praise, calling its authors 'two masters of the science'. Brande had taught for forty years, and the reviewer had 'had the good fortune to learn chemistry from the teachings of Dr Taylor, in the lecture-room of Guy's Hospital'. Taylor's own interests are stamped all over the book: from how to detect poison in wallpaper, to how to use chemistry in photography. They rejected chemical symbols; the *BMJ* reviewer was glad to see the back of 'those frightful formulae, symbols, and mystical language, which are enough to strike awe and repulsion into the mind of the ingenuous student'. The reviewer felt that Brande and Taylor had struck the right balance for the book's length – not too brief, not too exhaustive. Brande and Taylor had entirely left out illustrations to save space, testily commanding their readers to look up images of retort stands in catalogues. According to the reviewer in the *BMJ*, this had resulted in 'one thick and handy volume', which would otherwise have been 'two equally thick and inconvenient volumes'. Letheby appeared twice in the book, without any vitriol from Taylor. Perhaps his wrath had run its course, or perhaps Brande had toned down his friend's comments.

There is a studio portrait of Taylor and Brande, possibly taken at the time their joint book was published. [Plate 30] An enormous swagged curtain hangs down behind them, and white-haired Professor Brande stares away across the room. Taylor, his hair almost as thick and dark as it appears in the oil painting from his youth, looks directly into the camera. His expression is rather stern, perhaps the effect of sitting still for the exposure, or because he wished to project an aura of gravitas – or maybe he was unimpressed by the photographer's technique.

That same year, two elderly ladies from Taylor's circle died. 92-year-old Mary Walker had been the Taylor's housekeeper for many years. She left her money and possessions to her nieces, one of whom still worked for the Taylors. After Mary's death, Taylor had to swear to the truth of her will in Chancery. It must have made a pleasant change, after swearing to the evidence of so many violent deaths.

In August, Taylor's 90-year-old aunt, Dorothy Armsby, was buried near Taylor's parents in the churchyard at Northfleet. Not having any surviving children, Dorothy's will is full of nieces and nephews. Taylor's brother Silas and his children appear in the will, but Taylor and his daughter do not; Dorothy perhaps felt that Taylor did not need her help. The eccentric orthography and grammar of Dorothy's will say much for the education of women in the late eighteenth century, and one wonders what she made of her much-published nephew.

The Dead House

There are at least two instances recorded of Taylor being asked to estimate a time of death based on a body's temperature when it was found. Considering his general fascination with temperature, it was hardly surprising that he would eventually research how body temperature intersected with crime. At the end of October 1862, Samuel Gardner, a chimney sweep, was on trial for murdering his wife. The temperature of her body, and the advancement of rigor mortis, could have been interpreted in such a way as to incriminate the Gardners' servant, against whom there was no other suspicion. It was time that someone worked out a scheme to evaluate how the body cooled, and so Taylor drafted in Dr Samuel Wilks, a specialist in morbid anatomy, to help him. They worked in a hospital, after all; they had access to a plentiful supply of bodies.

From 21 January until 7 June 1863, Taylor and Wilks worked their way through 100 former patients of Guy's Hospital. They recorded the name and age, the cause and time of death, and the temperature of the body when it was taken to the hospital morgue, or, to use the unsentimental Victorian term, the 'Dead House'. They noted down how long after death the body had been removed there, and took up to three more temperature measurements.

The measurements were taken in a busy hospital, which may explain why the records are decidedly uneven. It wasn't until the thirty-first patient that they recorded the temperature of the ward where the patient died, and the thermometer was not placed in the same part of the body on all patients. On some, it was placed on the abdomen, whereas in cases where a post-mortem was performed, it was placed inside a body cavity. As other writers have pointed out, placing the thermometer on the skin left their readings vulnerable to draughts. There are notes recorded by some entries to say that the limbs were stiff or pliant, but not by enough cases for it to be a useful reference tool. The thoroughness of the other details is fascinating, however – the pneumonias, burns, tetanus, diseased kidneys, and vast variety of other causes that would rob a Victorian Londoner of their life. There is 35-year-old John Cooper, who is recorded as a 'negro', and a 17-year-old patient called Manuel Rajas; that they were at Guy's Hospital reminds us that in 1863, London was a global city.

Taylor and Wilks were not the first to record the cooling of the body after death, but they were the first to try to investigate it in such a way that it would be useful for medical jurisprudence. As ever, Taylor's wide knowledge of other cases fills the article, referring to earlier studies. He discussed the Gardner case, as well the Doige case, and even made reference to the Hopley case. 'Criminals sometimes unknowingly furnish important evidence in reference to the condition of the dead body.' Because Hopley had said 'the boy' was found cold and stiffening at about six o'clock in the morning, Hopley was pointing the finger at himself; the boy must have died just after Hopley was known to have beaten him the previous night. There is no mention of Taylor's personal connection to the victim, but it seems as if Taylor felt duty bound to mention poor Reginald's plight whenever the chance arose. It was the least he could do for his nephew.

This unhappy young woman

There were dozens of suspicious deaths that Taylor lent his analytical skills to during the 1860s. The deaths of Sophia Bragg, a landlord's wife in Slough, the Southern children in Earls Colne, Essex, and 14-year-old Jane Clarke of Westminster, might have looked like poisonings, but after investigation, Taylor proved that they had died of natural causes. In 1868, Taylor's analysis led to the conviction of Priscilla Biggadyke of Stickney, Lincolnshire, who was found guilty of poisoning her husband with arsenic. She was the first woman to be hanged within prison walls, after public executions had been brought to an end in Britain by an Act of Parliament that year. Toxicological cases followed the pattern that had passed before Taylor's eyes over and over again down the years: a sudden death, symptoms suggestive of poisoning, viscera in a jar, analyses in a laboratory, revelations at an inquest. But unusual cases still cropped up.

In May 1863, Taylor returned to the county of his birth, to investigate a murder in Herne, Kent. Richard Steed had been found dead, his skull smashed, after an argument with his former lodger, Alfred Eldridge. Taylor was given Eldridge's boots to examine and, with Dr Pavy, used their microscopes to match hair from the hobnails on the boots with samples of Steed's hair. They found strands of red wool in the coagulated blood on the leather of the boots. When they found out that Steed had worn a red woollen comforter, it was extra evidence to show that Eldridge – or someone wearing Eldridge's boots – had killed the unlucky Steed. Eldridge was hanged.

In November 1863, Elizabeth Waterer was found dead in a room at the Coachmaker's Arms in Guildford. Her soldier lover, Serjeant Joseph Mahaig, was lying on the bed beside her with his throat cut – but he was still alive. It was known that Elizabeth had bought rat poison, and, although the medical witnesses thought that she had been strangled, Taylor performed a toxicological analysis as a formality.

Elizabeth had bought Butler's Vermin Killer; when Mary May had bought it back in the 1840s, it had contained arsenic, but as arsenic sales were restricted, Butler's now contained strychnine instead. Taylor had not been able to find any strychnine in Elizabeth's body – Letheby no doubt curled his lip and tutted – but he said that, to the untrained eye, a death from strychnine could look like strangulation because it kills by its spasms suffocating the victim. There were marks on Elizabeth's throat, which the other medical witnesses thought were from strangulation, but Taylor attributed them to decomposition.

Elizabeth and her lover had written suicide notes, and these were irresponsibly reproduced in full by the newspapers. They were star-crossed lovers, who would be together in death. The inquest jury found that Elizabeth had committed suicide while in sound mind, and Mahaig had abetted her. At trial, the jury found Mahaig guilty as an accessory before the fact, but recommended him to mercy, on account of his good character and the circumstances around the crime. The judge sentenced him to death, saying, 'there could be no doubt that he had acted very ill to this unhappy young woman', and that his conduct had caused her such distress that 'she had resolved to destroy herself by taking poison, and thus put an end to the state of misery in which she found herself.'

There were cases on record of failed double suicides, where the surviving lover was found guilty of murder, and then pardoned. The Home Secretary and the judge corresponded. The judge did not agree with Taylor, perhaps his inability once again to find strychnine making for an unconvincing case. But even if the judge had believed Taylor, there was still the matter of Mahaig's possible coercion. The Home Secretary did not pardon Mahaig, but in deference to the jury's recommendation to mercy, he commuted the sentence to penal servitude for life, and Mahaig ended his days in Australia.

In 1864, a 9-year-old died in the Suffolk village of Wissett. The local surgeon thought that the girl had been poisoned, and an inquest was opened. Taylor examined her viscera, and found arsenic in her stomach and intestines, but it did not seem to have been administered orally. He analysed the child's scalp and found traces of arsenic and mercury. Her stepmother testified that she had put powder on the girl's scalp to treat an infestation of ringworm and vermin, and Taylor concluded that arsenic, when applied externally, could be absorbed by the body. It was known that it could be carried by the blood, and Taylor had written about it before. Here was tragic evidence of it in action. The coroner's jury decided that the parents hadn't meant to kill the child, but they did want to point out that the parents had neglected and ill-treated her.

The first railway murder

Later that year, Britain was rocked by its first murder on the railway. Well-to-do Thomas Briggs had been on his way home to Hackney from central London, when his body was found on the railway line. There was blood all over the interior of a train carriage, and on its wheels. The assumption was that Briggs had been fatally struck inside the carriage, and then his body had hit the wheel as it was flung out. Taylor was not consulted on whether or not the red liquid was blood, as this was referred to Letheby. But later in the investigation, Taylor analysed the lining of a man's coat, which bore marks suggesting that it had been used to wipe up blood. The fabric had been shoved up a chimney in a room where a man called Franz Müller had been lodging; he was the prime suspect. Taylor's analysis showed that it was human blood on the fabric.

It was possible that Briggs had been killed with his own walking cane, but there was only a small amount of blood on it, which could have come from being in the bloody carriage. The police later found a cane at the same lodging house where they had found the sleeve. It had a heavy lead finial, which could have administered a fatal blow. There was a substance on it that resembled dried blood, so it was sent to Taylor to analyse. He showed that it was not blood, but resin.

It was a sensational case involving a transatlantic chase, pawned watches, and the fashion in hats. But in finding human blood on the sleeve lining, Taylor had helped the police add a straw to the heap of evidence against Müller, who was eventually found guilty of Thomas Briggs's murder and hanged.

A wedding

In July 1865, there was a wedding. 21-year-old Edith, the Taylor's only surviving child, married Frederick John Methold, a 24-year-old clerk in the Common Pleas Office. The wedding took place at the church of St Stephen the Martyr, near the Taylors' home. Frederick's father was a Master of the Court of Common Pleas, as Edith's late uncle had been; Edith and Frederick might have been family friends for some time. Frederick's great-uncle was Nicholas Conyngham Tindal, the judge who helped formalise the McNaughton defence: 'not guilty by reason of insanity'.

St Stephen's vicar had filled out the marriage register, but unusually had left the box blank for the rank or profession of the bride's father, so that Taylor could complete it himself. He entered: 'MD FRS'. Taylor bought a house on Regent's Park Road for the young couple, and less than a year after the wedding, Taylor became a grandfather.

Seeing something done

After his daughter's wedding, Taylor dashed off to the Select Committee on the Chemists' and Druggists' Bill. Arsenic sales were restricted, but Taylor strongly felt that more needed to be done.

He warned that death might follow if shop staff were incompetent, and he presented the example of an assistant who supplied arsenic instead of calomel; it's likely he was referring to the Dore trial of 1848. He presented the case of a friend who had bought an ounce of tincture of rhubarb, and had been sold laudanum instead, narrowly escaping with his life. He related the shocking morning when 'upwards of 300 cases of poisoning by arsenic came before me at once from an industrial school near London'. An engineer had put some arsenic in the boiler to clean it, and, unaware, someone had added some of the water to the children's milk. Fortunately no one was killed, but when Taylor reported it to the Secretary of State, he was told that nothing could be done. There was a new danger from recently discovered nitro-benzole, which was used as almond flavouring, but had the unfortunate side effect of killing people. Taylor was asked what he would do – did he wish to fetter commerce? 'I am more impressed with the necessity of seeing something done to prevent these deaths from occurring,' he tersely replied.

Taylor's concerns for restricting access to poisons had never ceased. The year before, he had made a recommendation that bottles containing poisons should have an external texture that would immediately alert the unwary to their contents. He, and the Coroner of Liverpool, particularly suggested using Mr Thonger's patent labels, which were made of sandpaper, but domestic poison bottles did not have to carry ridges or grooves until the twentieth century.

The Pharmacy Act was passed in 1868. Finally, strychnine, prussic acid, ergot, opium, and poppy derivatives were regulated. Chemists were unhappy, as laudanum made up a lot of their business, but for those who saw the misery of laudanum addiction, there was little pity for them.

Rare qualities as a compiler

Taylor's *A Manual of Medical Jurisprudence* was published for many years, even after his death, as a convenient, small textbook. Towards the end of 1865, twenty years after the *Manual* had first hit the shelves, Taylor added an extra book, a 1,186-page volume called *The Principles and Practice of Medical Jurisprudence*. For the first time, Taylor had space to include illustrations – the skeletal remains of the Waterloo Bridge Mystery, magnified crystalline forms of poisons, fibres of fabric and hair, blood spatter, leaves of poisonous plants from photogenic drawings, body parts, magnified bodily fluids, the different facial expressions of the mad. Extra space meant that he could expand on sections that appeared in the *Manual*, and include new subjects.

The Lancet welcomed the book, praising it especially for the introduction, which laid down 'general principles and rules, which are of equal importance to all skilled medical witnesses, and which none ought to disregard'. *The Lancet* pointed out to its readers that, 'of all the professions, ours is that which is most frequently called into the witness box,' and Taylor's book was the perfect primer for all medical men. Despite their praise, they regretted defects that arose 'out of personal bent of character and strong feelings engendered by past conflicts'. In places, Taylor used his book 'to pillory experts who have been opposed to him in various cases'. They condemned 'the frequent allusions to the Palmer and Smethurst cases, and the angry denunciations of the opposing witnesses, as the extreme of bad taste'. But even so, they recommended the book, and hailed Taylor's 'rare qualities as a compiler', his accomplishments as a chemist, his lawyer-like acuteness, and his ingenuity in how he laid out his arguments.

The Principles and Practice of Medical Jurisprudence would be published for more than 100 years, material removed and added to make it relevant to its era, until its final edition in 1984.

A test for blood?

When analysing bloodstains, Taylor used a magnifying glass and a microscope, but in the 1860s, a test was formulated using guaiacum, a resin from a West Indian tree. There had been suggestions for chemical tests for blood before, but Taylor hadn't rated them very highly. The test for blood is based on the resin turning a rich sapphire blue when oxidised, which it will also do when a mixture of blood and hydrogen peroxide is added. Unfortunately there are other substances, such as pus, which can give the same result, and Taylor corresponded for several years with Dr Day in Geelong, Australia, who refined the test. In careful hands, ensuring the guaiacum mixture was fresh, and that anything that could create a false positive was ruled out, it could help forensic scientists identify blood. But it wasn't a perfect solution.

In 1867, the case of John Wiggins was referred to Taylor and his colleague Wilks. Wiggins was indicted for the murder of Agnes Oaks, who had been found dead with her throat cut. A small wound was found on Wiggins's neck. The prosecution alleged that Wiggins had killed Agnes and then cut his own throat to deflect suspicion from himself, whereas the defence alleged that Agnes had tried to kill Wiggins, then turned the knife on herself. Taylor used a combination of evidence, including that of blood, to try to get to the truth of the matter.

Taylor examined Wiggins's clothes, and found blood on them, particularly on his stockings. It seemed as if he had walked across a bloodied floor without any shoes on. There was his neckerchief, which had apparently been cut when the wound to his neck was made. Taylor was sceptical; he argued that the cloth had been removed and then cut. Firstly, there was no blood around the cut in the fabric, as would be expected if the cut had resulted in a gash in the skin underneath. Secondly, the neckerchief was tightly folded and the knife had passed through sixteen layers of cotton. Taylor claimed that the knife must have been thrust with great force to cut through that much fabric, and that the wound in the neck was too small.

Agnes's wound and the position in which the body was found mitigated against the idea of suicide. Taylor pointed out that the slash across her neck started with a deep stab that penetrated to the spine. It might not have killed her straight away, but it would have prevented her from moving. How did she manage to continue the cut across her throat? How did the knife end up on the other side of the room from the body? How did the body end up under a chair, lying at an angle that made it impossible for her to cut her own throat?

Taylor found himself agreeing with the prosecution, and his evidence helped to convict Wiggins of murder. When Taylor wrote about the case, he emphasised that the crime scene provided vital clues that would suggest whether a death was murder or suicide.

Sweet Fanny Adams

The expression 'sweet FA' is an abbreviation of strong language; euphemistically, people replace it with 'sweet Fanny Adams' to mean 'nothing at all'. Fanny was a child who was murdered and mutilated in August 1867, in Alton, Hampshire. It was so gruesome that sailors in the Navy called their tinned meat 'Fanny Adams'. Suspicion fell on a man called Frederick Baker, a 29-year-old solicitor's clerk, who had been identified by other girls who had been playing with Fanny. They said the man had given them ha'pennies to go home, and he had carried Fanny away.

Taylor was given knives and clothes that belonged to Baker, and a stone that had been found near the scene. There was a red substance on them, and after a month of careful analysis, Taylor was able to say that it was human blood. But at the trial, Taylor was cautious. Baker's penknife was produced and Taylor said 'he found it impossible to believe that the knife had been used in the manner described.' He didn't think it was large or sharp enough. Neither did there seem to be much blood on the clothing. 'Blood enough to fill a thimble would make a great show,' but what Taylor saw could be accounted for by a nosebleed. When asked at the trial how long it would have taken to dismember and disembowel a child, Taylor replied that 'even an inexperienced person need not be more than half an hour if he had proper weapons – a sharp knife or even a table knife.' The responsible Victorian press ensured that this information was widely circulated so that all would-be murderers had Taylor's instructions to hand.

Baker's defence counsel asked Taylor about 'homicidal mania'. As there was mental illness in Baker's family, this was one line of their defence, other than Baker having been misidentified. Whilst Taylor said that homicidal mania was an impulsive killing, it was usually 'of some loved object', and there would be 'no attempt to conceal the crime'. These words might have led the jury to think that Baker did not suffer from 'homicidal mania', as Fanny was a stranger to him, and he had returned to the scene to try to bury some of her remains.

Other evidence against Baker was strong: he had been seen in the field around the time the girls had been playing, and a colleague from Baker's office knew he had been absent for some of that day. Baker had been drunk when he met him later, talking about going away once rumour was afloat that he had killed Fanny. His diary was used in evidence, where he had written 'killed a young girl; it was fine and hot', but it may not have been an admission of guilt but an observation that a young girl had been killed. 'Fine and hot' was his usual way in his diary of describing the weather. Public feeling was so strong against Baker that the defence said 'were his client at liberty at that moment, the people themselves would commit murder by tearing him to pieces.' The jury found him guilty.

Taylor was very well known, and his portrait took pride of place on the front cover of *The Illustrated Police News* when they reported the trial. It was 'from a photograph', they admitted, although they had added in some details to make it look as if he was in the witness box. It was a flattering rendition, making Taylor appear heroic and younger than his 60 years. At least they were honest about doctoring their images, unlike the *Illustrated Times* a decade earlier.

Before his execution, Baker confessed. Even though Taylor had found it hard to believe that the penknife was the murder weapon, Baker insisted that it was. He claimed that he had been 'maddened by drink', and had done it 'in an unguarded hour, and not with malice aforethought'. It is possible that he had hoped for a reprieve based on insanity, but Calcraft hanged him on Christmas Eve, *The Illustrated Police News* devoting another front page to the case. They tastefully looped Calcraft's portrait in a frame of rope.

Eminent Medical Men

In 1868, a five-page biography of Taylor appeared in *Photographs of Eminent Medical Men, of All Countries;* despite the book's title, they are almost all English. Much of the information seems to have been provided by Taylor, as it narrates his early life in detail. Surrounding him are the great and good of nineteenth-century medicine, men who made inroads in disease and surgery. Taylor appears to be out on a limb, but that he appears in the book demonstrates how much his chosen specialism had gained in status over the years, in large part due to his own efforts. The biography acknowledged his role: 'It is given to few men to follow for nearly fifty years a study almost new and untried, and, to a still smaller number, is it permitted from such long and patient study, to educe results so valuable and important as, in the now comprehensive science of Forensic Medicine, Dr Taylor has attained.'

Each man posed for his photograph at the London studio of Ernest Edwards, on Baker Street. Nunneley appears in the book just before Taylor, his enormous whiskers horripilating in impressive fashion. William Augustus Guy is a few pages after Taylor, slouched suavely in the armchair that Taylor had been leaning on for his picture. Another image attributed to Edwards, taken the same year but not used in the book, shows Taylor seated, holding a photograph. [Plate 31] It's as if the two men had discussed the art of photography, Taylor passing on his secrets to the younger man.

Scandal

Aside from all the criminal cases Taylor worked on during the 1860s, he analysed water from the Aldgate pump in London, and from the river Lea in Luton. He wrote Professor Brande's obituary after his friend's death in 1866, and prepared a new edition of *Chemistry* alone. He even found himself on the same side of the courtroom as Letheby during a patenting case for magenta dye; Taylor's erstwhile sidekick Odling was on the opposing team. Tenacious as ever, he was still agitating for change in the way that coroner's inquests were performed. In 1867, the Food Committee tasked him with various analyses of meat. He examined a tin of meat that had been sealed in 1826, and made the unsurprising observation that 'it could not be expected that the contents of a tin case, whatever process of preservation may have been used, would be fit for food after the lapse of forty years.' When cattle plague hit Derbyshire, there was a rumour that Taylor would be sent a cow's stomach to investigate, and he was quoted in a long pamphlet advert for Spencer's Patent Magnet System of Purifying Water.

It seemed as if, wherever science met everyday life, Taylor would be found. Even in people's private lives. Being gay or transgender in the nineteenth century was difficult. Up until the 1860s, sodomy had carried the death sentence, but by 1870 could be punished by a long sentence of imprisonment with hard labour. Taylor's books on medical jurisprudence carried sections on sodomy, because it was illegal between same-sex and heterosexual couples, providing grounds for divorce.

Frederick Park and Ernest Boulton – or Fanny and Stella – were two young men who were 'female impersonators' on and off stage. There was evidence to suggest that they engaged in prostitution, and angelic-looking Stella was the 'wife' of Lord Arthur Pelham-Clinton. When Park and Boulton were arrested in April 1870, after lewd behaviour at a theatre, it was headline news.

The trouble was that causing a public nuisance was a minor crime compared to the far greater allegation of sodomy: there needed to be solid evidence to prove the graver offence. After their arrest, Park and Boulton were taken to Bow Street police station and were intimately examined by a police surgeon. He had been one of Taylor's students at Guy's Hospital and was fascinated by the 1833 case of Eliza Edwards.

In June, *The Lancet* reported that Park and Boulton's trial was fast approaching, and they expressed concern that there might be difficulty 'lest a new scandal may be caused by directly conflicting medical testimony'. *The Lancet* believed that British medical men didn't have adequate first-hand experience of sodomy; it just wasn't very British. They asked that medical experts be sent from abroad. 'There are physicians in Paris, in Spain, and in Constantinople, who must, unfortunately, have seen very much of the results of a certain hideous vice.' Nevertheless, Taylor was one of the experts who was summoned to examine Park and Boulton in a grotty cell in Newgate prison, even though he rarely seems to have examined living people, being far more at home with the disembodied organs of the dead. But Taylor had read about the subject, and written about it in books; as one of the country's leading medical jurists, he was called in by the government, roped into a case that was causing scandal across the country.

A curious fact emerged that, while Boulton had surgery the year before, it was Taylor who had administered the chloroform. One wonders just how many people Taylor acted as anaesthetist for, but as a chemist who was an apothecary, surgeon, and physician combined, patients were – perhaps – in safe hands.

Just as *The Lancet* predicted, the medical gentlemen couldn't come to a consensus. Taylor was one of the men who did not believe there was sufficient medical evidence to support the charge against Boulton and Park. The lack of agreement put a dent in the case, and without medical proof of the offence, they could be presented as two silly boys who had gadded about ill-advisedly in women's clothing. They were acquitted.

Of the vast quantity of literature generated about Boulton and Park was a two-penny pamphlet of questionable quality called *Life and Examination of the Would-Be Ladies*. It described the pair as 'England's gems in petticoats'. Taylor owned a copy, as part of his huge collection of medico-legal pamphlets that he had acquired over the years.

It was to be one of the last famous cases of his career.

Chapter 17

The Eminent Opinion of Professor Taylor
1870–80

An unretiring retirement

Time went on apace. Taylor resigned from his chemistry lectureship at Guy's in September 1870, passing the baton to Dr Heinrich Debus and Dr Thomas Stevenson. Stevenson was thirty-two years younger than Taylor. Just as Taylor had won the anatomy prize all those years earlier, Stevenson had won the anatomy medal, and he had also won the hospital's gold medal in forensic medicine, which had only come to exist through Taylor's labours.

Taylor retired from the witness box, so in the 1870s it was Stevenson who went the rounds of inquests and trials in his place. In 1872, when Ellen Kittell stood accused of poisoning her husband's first wife, it was Stevenson who made the journey to north-east Essex. Taylor knew the area only too well from the trials of Mary May and Hannah Southgate, which had taken place when Stevenson was a child.

When the census was taken in 1871, it found Taylor and his wife Caroline at home at St James's Terrace. Caroline had a lady's maid, and as well as four other domestic servants, the Taylors also had a coachman. In the past, Taylor presumably drove himself about town, but perhaps age was against that now. His brother Silas, meanwhile, was going back and forth to Dieppe in northern France, continuing his work as a merchant. He appears on the census with his wife's nephew, who was a manufacturing chemist. Silas was never as financially successful as his brother; when he died in Dieppe in 1898, aged nearly 90, he left only £68 in his will.

Although Stevenson was taking on Taylor's criminal cases, the older man was still in demand. In 1871, 'the eminent opinion of Professor Taylor' was sought on a possible rabies death in Derbyshire. He supplied an analysis of the water supplied by the West Middlesex Water Works Company, and there was more meat for him to test for the Food Committee. In 1873, he contributed to the *BMJ*'s campaign against homeopathy. His submission was 'of an extremely decisive and weighty character'. He could not resist referring back fourteen years to the Smethurst case, when among the medicines he and Odling had analysed were 'sixty-four small tubes of homeopathic globules … including, as would appear from the attached labels, every variety of mineral and organic poisons and medicines'. By coincidence, Smethurst died that year.

Taylor appeared in another photographic compendium in 1873, *The Medical Profession in All Countries*, which once again, despite its title, almost entirely featured Englishmen. [Plate 32] Taylor's biography mentioned that *On Poisons* had been translated into Dutch and German; in 1879, *Principles and Practice of Medical Jurisprudence* would be published much further afield, when it was translated into Japanese. Considering his travels as a youth, and how his reading was international, it seems only apt that his writing spread as far as it did. At the beginning of his career, *The Lancet* had mocked Taylor's youth by calling him 'beardless'; now nearing seventy, his whiskers were few compared to his grizzly faced peers who appear elsewhere in the book.

When *The Principles and Practice of Medical Jurisprudence* went into its second edition in 1873, it spilled out into two volumes, almost 1,600 pages long in total, with nearly 200 illustrations. The prodigious size of Taylor's book was not popular with all, even though the condensed *Manual* was still being published. A doctor in Liverpool told *The Lancet* that a student had asked him what was the best book on medical jurisprudence 'for his study, and … how much was considered essential for him to become acquainted with'. The doctor had replied, 'I believe the essentials of the subject may be put into about fifty pages of octavo.' Taylor must have been horrified when he read this, especially when the doctor singled him out for criticism. 'I naturally hesitated to recommend Dr Taylor's work on this subject, which has of late years assumed such gigantic proportions.'

Among the illustrations was a crime scene diagram showing the 1862 murder of Mrs Gardner, one of the cases that had led to Taylor's research into the cooling of the body after death. The diagram was based on a drawing made on the spot by Dr Sequeira, who had been called in on the discovery of the body. Mrs Gardner had been found with her throat slashed, and a knife in her hand, but Sequiera's crime scene diagram, along with his sketch of the slashes to Mrs Gardner's hands, helped to show that she had not committed suicide.

Taylor was keen to follow cases up. In the first edition, he had written about the still-unsolved 1860 child murder at Road Hill House in Somerset. He veered away from the strictly medical to talk about the fact that the killer had to have been in the house – this aspect of the murder would partly inspire the 'locked room' mystery of crime fiction. But Taylor's first edition had gone to print in 1865 almost at the same time as Constance Kent, the victim's teenage sister, had confessed to the murder. So, eight years later, Taylor returned to the case in the second edition. Constance had been 'a girl only sixteen years of age, [who] showed an amount of cunning, in the perpetration and concealment of this act of murder, rarely met with among old and experienced criminals'. From the wounds on the child victim, 'it was not probable that the dress of the person inflicting them could have escaped being stained.' Constance had fooled almost everyone by destroying her bloodied nightdress. Taylor's feelings about the failure of the murder investigation is clear when he comments about the nightdresses not having been accounted for at once: 'A heavy load of suspicion would have been removed from several innocent persons, and a crime like this would not have remained concealed for five years.'

Taylor extended the section on 'Concealed Sex', which in the 1865 edition had only featured Eliza Edwards. Dr James Barry had died just as that edition had been published, and the discovery was made that this former student of the Edinburgh Medical School and Guy's Hospital had been assigned female at birth, and had lived as a man for most of his life.

Christison, whose opinion and support Taylor had valued over the years, had waged a determined battle, trying to prevent women, 'The Edinburgh Seven' led by Sophia Jex-Blake, from studying medicine at Edinburgh. He had blocked them at every step, using his enormous influence and his powerful connections, even claiming that Queen Victoria agreed that women should not practise medicine. Taylor was not of the same mind as Christison; after all, his wife helped compile his books. Taylor was now so well established that he could afford to be cheeky. At the end of his piece about Dr Barry, Taylor wrote, 'With such a successful precedent before them the Examining Board of Edinburgh are hardly justified in excluding women from professional study and examination.'

He might have felt a twinge of professional jealousy. In 1871, Christison had been granted a baronetcy by Queen Victoria; properly, he was now Sir Robert. This was not necessarily for his medical jurisprudence work; he had not occupied the chair at Edinburgh since 1832, having been Professor of Materia Medica ever since. Even though Christison had continued to work on criminal investigations, he was one of the Queen's physicians in Scotland, and he was an important figure at the University of Edinburgh and in the city itself. Taylor and Christison had both applied science and medicine to solving crime, but their careers had in the end panned out quite differently.

As it happened, Thomas Stevenson would become Sir Thomas, when he was knighted in 1908 for investigating suspicious deaths for the Home Office. Stevenson was associated with the famous late nineteenth-century poisoning trials of Florence Maybrick, Thomas Neill Cream, and Adelaide Bartlett. By then, medical jurisprudence had assumed a position of authority that it was reaching towards during Taylor's lifetime but hadn't quite attained.

Beyond the auspices of his profession, Taylor kept himself to himself; he wasn't a big noise in scientific societies, so he was beneath the notice of those who bestow honours. During his career, Taylor had done his best to annoy the establishment, pushing for regulations that would have impinged on enterprise; there were perhaps people in high positions who did not want to see him honoured.

The West Haddon Tragedy

The dangers of drawing inaccurate conclusions from an analysis became abundantly clear after the West Haddon Tragedy in 1873. Elderly Mrs Gulliver had been nursed by her niece, Mrs Waters, over the last two days of her life. No one else fed her, or gave her medicines. Her symptoms pointed to heart disease, and the surgeon who attended her gave heart disease as the cause of death. But her end seemed to have come on suddenly, and rumour spread that she had been poisoned.

She was exhumed a month after her death, and in December 1873, a solicitor applied to Taylor to make a toxicological analysis. The viscera arrived at Guy's, too late for Taylor to inform them that he had retired 'from this branch of practice'. He referred them to a chemistry professor, who was also an MD, but he wasn't in when the police called. As the case was deemed urgent, they went to another man: Julian Rodgers, whom Taylor had crossed swords with before.

Rodgers claimed that morphine had killed Mrs Gulliver 'in traces' – unweighable amounts, which other experts said might have come from cough lozenges. Rodgers claimed that after death, Mrs Gulliver's body had abnormally retained its heat. He apparently based this on the surgeon finding the bedroom window open, as if a guilty person had been trying to cool the room. Rodgers claimed that this abnormal retention of body heat was from morphine: it was murder. It was supposed that Mrs Waters had a pecuniary interest in her aunt's death, and the inquest jury returned an open verdict; someone unknown had murdered Mrs Gulliver. It could only have been Mrs Waters, and wild with terror, she destroyed herself with strychnine.

There was a huge amount of comment on the case. Rodgers would not back down, even when Taylor said that in his research with Wilks into body temperatures after death, people who died of heart disease did not cool rapidly; the evidence of the open window did not exclusively point to poisoning. Christison weighed in with his opinion, as did Wilks, and his colleague Walter Moxon. They all felt that the traces were insufficient to prove death by poisoning, particularly as there were no symptoms, and the natural causes of heart disease accounted for the death.

Taylor wrote an article for *Guy's Hospital Reports* on the case. He may have felt a twinge of guilt that he had not made the analysis when the viscera had been in his hands. Coroners' inquests were inadequate, Taylor said: 'The whole system requires reformation, and until this takes place, we must be prepared to meet with such tragical cases as that which is described in this paper.'

His very last article for *Guy's Hospital Reports* was on using tattoos for identification. This followed the sensational case of The Titchborne Claimant, who arrived from Australia, claiming to be the missing heir to a fortune. After a trial lasting 188 days, the Claimant was found guilty of perjury: he wasn't Sir Roger Charles Tichborne at all, but 'Arthur Orton, the son of a butcher at Wapping'. Taylor's article explored whether tattoos could disappear over time, or be removed, and whether a competent medical professional could discover where they had once been on the body.

Destruction in every room

'At five o'clock yesterday morning, just as day was breaking, all London was startled by a tremendous Explosion.' So wrote *The Times* on 3 October 1874. The Taylor household was woken by a massive bang, and the house shook. Taylor wrote to *The Lancet* that 'in an instant a heavy shower of broken plate-glass from the windows covered the bed in which I was lying, as well as the floor of the room.' A near neighbour of Taylor's also wrote to *The Lancet*; 'the screams of my family ... were so piercing, that I was quite prepared to find them dead or dying.' Taylor's hands and feet had been cut by the glass, and his clothes beside the bed were covered in it. His first thought was that the house had been struck by lightning, but 'on looking out of the open space left by the destruction of the windows' he could see that there was no storm. A few seconds later, his bedroom was filled with thick smoke, and the 'strong smell of gunpowder'.

He managed to get dressed and examine the house; 'I found destruction in every room.' All of the windows were shattered, the furniture had moved, the doors had been forced open, and the 'shutters rent from their hinges'. Taylor commented that it looked like the house had been hit by an earthquake. They soon heard that it had been caused by an explosion of gunpowder on a barge as it passed under the North Gate bridge on the Regent's canal, very close to St James's Terrace.

Taylor's letter to *The Lancet* described the damage to his house in enough detail that we find ourselves on a tour of his home. The dining room had a bird in a cage, and heavy mahogany chairs, and his study faced north, with an ornamental sash window and stained glass. But that wasn't the purpose of his letter. He wanted to describe the damage caused by the explosion, and to complain about the risk to public safety caused by the transportation of gunpowder and petrol through residential areas on a steam tug boat.

'I have seen this steam-tug from my bedroom windows during the dark winter nights with sparks and flame issuing from its funnel, appearing like a fiery meteor amidst the trees.' The transportation of gunpowder and its risk to public safety had been discussed by Parliament, but the usual decision had been made: 'it was considered that legislation would paralyse the trade; therefore nothing was done.' The result was at least £200,000 worth of damage in the streets around Regent's Park, and the liberal dosing of chloral hydrate to sooth the residents' fractured nerves.

On Poisons

In 1875, Taylor produced the third edition of *On Poisons*; it had not been updated for sixteen years. For the first time, it contained illustrations – over a hundred, mainly showing the leaves of poisonous plants. In the preface, Taylor explained that he had added many things, and removed some others. Readers might have been forgiven for wondering why no mention at all was made of the Smethurst case, when he had been so fond of alluding to it at every opportunity. He mentioned the Palmer case a couple of times in passing, his bitterness still evident, but he named no names and his grumpy digs were but fleeting: age had mellowed him.

All our days are gone

On 8 February 1876, Taylor's wife died. They had been married for nearly forty-two years. An announcement was published in the *Pall Mall Gazette*, and Caroline was buried in Highgate Cemetery, 3 miles north of her home.

On the one hand, it had been a highly respectable marriage, living on Regent's Park, their neighbours wealthy barristers and merchants. But it had been dominated by Taylor's profession. Caroline had helped him to compile his books, without her name ever appearing on the title page. When visitors were announced, they might have been ladies come for tea, or they might have been policemen bearing bloodied trousers, or surgeons carrying specimen jars. She and her daughter had sat at the dining table, Taylor's place empty while he travelled the length of the country for inquests and trials. She had read angry criticism of him in the press, and they had borne a painful silence when the death of her nephew was the fascination of the country.

And now, Taylor was without her.

Though men be strong

There was still work to be done. Taylor turned 70 in December 1876; in the same year that he lost his wife, he co-edited *Arnott's Elements of Physics*, and contributed to a report on the Cruelty of Animals Bill, focussing on vivisection.

In 1877, Taylor was one of a handful of experts consulted on The Penge Mystery. Harriet Staunton had starved to death, apparently by the neglect of her husband. It gripped the public, with scandalous details about the Stauntons' private life. Heiress Harriet had what we would now call learning difficulties. Her husband had misled her, all the while carrying on with another woman, and Harriet had borne him a child. She was, it seemed, left to starve in the room of a rundown cottage while her husband, his girlfriend, and two other accomplices waited for her to die so that they could take the poor woman's money.

The accused women melodramatically shrieked and fainted during the trial, and all four suspects were sentenced to death, but there was a question mark over it. Had Harriet really been starved, or had she been suffering from tuberculosis? Taylor commented on the medical evidence given at the trial, after sentence of death had been passed; Taylor's opinion was still trusted and valued in cases of great difficulty. Three of the prisoners' sentences were commuted to penal servitude, and one was freed.

In light of the Gulliver case and The Penge Mystery, it was all the more important that coroners' inquests were reformed. In early 1878, the *BMJ* published Taylor's long, detailed recommendations for the Parliamentary Bills Committee, inviting comment from its readers. Taylor had had more experience than most expert witnesses over his long career, and knew the good and bad of the system.

Take thy plague away from me

Taylor had rewritten his will less than a month after he was widowed, for the sole benefit of his daughter, her husband, and her children. He set up a trust fund for Edith, and he bequeathed his three houses, including a cottage in Northfleet – Taylor was still a Kent boy at heart – to his son-in-law. He bequeathed his son-in-law £2,500 and all his household goods, saying he had 'perfect confidence in his uprightness and integrity'. But a codicil added a few days later, possibly on his son-in-law's urging, had Taylor explain what he meant. 'I do not by such expression mean to impose on the said Frederick John Methold any trust or obligation enforced in law or equity but simply to express my feeling that he will do what is right with the property I have given him.' Taylor's own words fought with the rigidity of legal language, just as they had on numerous occasions during his career. He still had interests in Kent, besides the cottage in Northfleet, as one of the witnesses to his will was a solicitor from Gravesend.

Taylor was extremely wealthy, his personal estate being about £60,000. It's unlikely to have derived entirely from his work. Even though he charged five guineas for an analysis, he would have had to perform a huge number to accumulate such a sum. It seems to have partly come from private investments, inherited wealth, and his wife's income.

He retired from his post at Guy's in 1878, Stevenson becoming Professor of Medical Jurisprudence and Toxicology. At the end of 1879, William Calcraft, the hangman who had dispatched many of the people found guilty on Taylor's evidence, died. In early 1880, Taylor was one of three men on a sub-committee to the government to assist the Home Secretary with the Coroner's Bill.

At the same time, Taylor corresponded with photographer John Werge, lending him some of his early photographs for a lecture. Werge paid Taylor a visit to return the samples, later describing the 73-year-old as a man of 'remarkable energy and versatility'.

In early May 1880, Taylor made an alteration to his will, changing one of his executors to John Bonham Croft, a student at Cambridge. Croft was only 23, but his father was a physician who lived not far from Taylor, so he may have been a family friend.

The timing was important as Taylor evidently realised that he did not have long to live. Despite his medical friends rallying round, there was little they could do to halt his heart disease. The man who had stood shoulder to shoulder with mortality throughout his professional life, examining the dead for clues, and sometimes sending the living to their doom, was now on the brink of eternity himself.

Alfred Swaine Taylor died on 27 May 1880. He had asked that his funeral would cost no more than £30, and he was buried in the same grave as his wife in Highgate, under a slab of pink granite.

The Times's obituary detailed his professional achievements; his 'name is so well known to our readers in connexion with poisoning cases.' The *BMJ*'s obituary was written by someone who had known Taylor well. It was a warm, affectionate piece, which explained 'his ardent and sincere love of truth', and how tempted he had been during his decade-long retirement to return to the witness box. It gave an insight into how he wrote his books, with his diaries and carefully catalogued and indexed commonplace books; it was the only obituary to give Caroline credit. Taylor's faith was alluded to; 'when speaking of the wonders of Nature' he had often been heard to say 'that it was harder to be an infidel than a believer.' This may indicate that Taylor was not entirely comfortable with Darwinism. The *BMJ* revealed that for many years, Taylor had been supplying them with anonymous editorial on medico-legal matters; he was clearly rather proud of this as 'it was a fact which he himself frequently mentioned'. They listed his famous cases, 'such as those of Palmer, Smethurst, and Tawell', even though Taylor had only been involved in the Tawell trial because his book was used in the courtroom, and the obituary incorrectly claimed that he was the first lecturer of medical jurisprudence in England. These errors have been repeated ever since.

The Medical Times and Gazette began their obituary grandly enough. In Taylor's death there had 'passed away one of those men who are representative of an important branch of medical science', one of 'the brightest glories of the great school of medicine at Guy's, and the foremost representative of his department of medicine in the country'. Whoever wrote the obituary knew Taylor, or had seen him in the flesh. He 'was of a tall and imposing figure, gracious to friends and bitter to foes, and, as the lawyers found, a superb witness, not to be shaken by any light wind of doctrine.' They praised his work on bloodstain analysis, the Waterloo

Bridge Mystery, and the case of Thomas Drory, then repeated the *BMJ*'s mistake, attributing the Tawell case to him.

'But Taylor was not always right,' they said, and in detail, they described the paternity case that he had lost, and the Palmer and Smethurst trials. 'So strong was the influence of [the Palmer case] on Taylor's mind that it gave him a kind of mental twist, and he could never after recur to the subject without pain and irritation.' The writer was extremely critical of Taylor's work on the Wilson case, where 'by common consent, Taylor was seen at his worst.' But the critical obituary ended by praising Taylor for his books, and the work he had done to ensure that expert witnesses were adequately paid.

He was fondly remembered in the Harveian Oration, delivered at the Royal College of Physicians. 'Engaged as he was for many years (indeed, for nearly half a century) in the most difficult inquiries connected with medical jurisprudence, it is the highest praise to say of him that there is no instance where his services were not for the vindication of truth and justice.' Those who knew him would feel the loss of 'an instructive companion and a genial friend'. Who, they asked, would replace him?

Some years later, 'TS' – presumably Thomas Stevenson – wrote a brief profile on Taylor in the *Medico-Legal Journal*, a periodical that Taylor would have eagerly filled the pages of, but it only began publication after his death. Stevenson made mistakes, attributing the 'Terwell' trial to him, and said that Palmer had killed Cook with prussic acid. Stevenson ended his piece by saying, 'Those who knew him and saw how he was respected in private life, would scarcely suspect how keen and sarcastic he could be to foes.' Stevenson was doing some housekeeping for the sake of his mentor's reputation; Taylor's sniping sarcasm is evident to us even all these years later.

In 1882, Taylor's daughter and her husband paid for a stained-glass window in the church of St Stephen the Martyr to commemorate her parents. The church no longer exists; it was damaged during the Blitz and was eventually demolished. St James's Terrace was knocked down and replaced in the 1930s by a block of flats called St James's Close, but an echo remains in the road behind, called St James's Terrace Mews. Although Taylor's old house on what was Cambridge Place still stands, it does not have a blue plaque. There is, however, a blue plaque just around the corner on Albany Street, commemorating journalist Henry Mayhew for founding *Punch* magazine. He had got Taylor into a spot of bother during the Palmer case; no doubt the shade of the professor would have something sarcastic to say about that.

The most famous name

Taylor's legacy lived on. Most immediately was the contribution he had made before his death to what would become the Coroners Act of 1887. Although *On*

Poisons saw no posthumous editions, *Taylor's Principals and Practice of Medical Jurisprudence* was taken up and edited by generation after generation of experts, beginning with Stevenson. In Stevenson's preface to the third edition in 1883, he wrote that preparing it had been 'a labour of love'. Stevenson had felt it necessary to tone down Taylor's language about medical witnesses in the courtroom. 'However truthfully the author's lively descriptions may have represented scenes in Court between witnesses and barristers, as they appeared to his mind … it is rare indeed, nowadays, that a medical witness, when honestly endeavouring to state the truth, receives anything but courteous treatment at the hands of counsel.' So sounded the death knell for Taylor's tintinnaculatio.

A review of Nigel Morland's *An Outline of Scientific Criminology* in 1951 mentions Taylor; his books were 'the standard that has systematized the science in the English-speaking language'. Morland used Taylor as his authority for his chapter on medical jurisprudence, but the reviewer decided that by 1951, Taylor was 'ante-dated, if not antiquated'.

As each edition of *Taylor's Principles and Practice of Medical Jurisprudence* rolled by, less of Taylor could be seen. By the thirteenth and final edition in 1984, Taylor's only appearance, other than in the title, was his portrait at the front of the book as a young man, and a brief mention in a section on the history of medico-legal systems. Taylor, they claimed, was 'probably the most famous name in English legal medicine'. Even so, the Waterloo Bridge Mystery had no place in a world of finger-printing, CCTV and blood-type analysis. Toxicology in the modern world rarely involves such quaint poisons as strychnine and arsenic, when drug overdoses and the pollutants belched by factories are far more commonplace problems. Taylor's world was irrelevant to that of modern forensic medicine, and he was at risk of being forgotten.

I dabble with poisons

Wilkie Collins and Charles Dickens had been fascinated by Taylor's work; Collins especially had seen how it could be applied to fiction. But other authors would follow.

In 1876, Arthur Conan Doyle began his studies at the Edinburgh School of Medicine, where the teaching of medical jurisprudence had begun in Britain a century before. Three years later, a letter from 'ACD' appeared in the *BMJ*, explaining some self-experimentation with the tincture of a plant extract, gelseminum. It was a treatment for headache, but it was known to be poisonous, and in his letter, 'ACD' wrote about the physical effects he experienced on gradually increasing the dose. As Taylor was supplying editorial to the *BMJ* at the time, it's highly possible that he had read Conan Doyle's letter.

Conan Doyle said that it was Joseph Bell, one of his tutors, who was the main model for Sherlock Holmes, as well as Poe's detective Dupin. Curious visitors to the Christison family grave in Edinburgh will be furnished with a clue that will lead them to the interesting discovery that Robert Christison's granddaughter married Joseph Bell's first cousin. But that's not the only connection between Conan Doyle and nineteenth-century forensic scientists.

In the very first Sherlock Holmes story, *A Study in Scarlet*, published in 1887, Stamford tells Dr Watson that Holmes might well give his friend – and himself – 'a little pinch of the latest vegetable alkaloid, not out of malevolence, you understand, but simply out of a spirit of enquiry'. Conan Doyle had done this himself, but recall that Christison had put arsenic on his own tongue to find out if it had a flavour. An unambiguous reference to Christison appears when Stamford tells Watson that Holmes has been 'beating the subjects in the dissecting-rooms with a stick', just as Christison had done during the Burke and Hare trial in 1828.

Stamford says that Holmes is 'working at the chemical laboratory up at the hospital'. He is 'well-up in anatomy, and he is a first-class chemist'. When Watson first meets Holmes, he has just made 'the most practical medico-legal discovery for years'. And what might this be? 'An infallible test for blood stains,' Holmes declares. Holmes puts a plaster over his finger where he has pricked it to draw blood, explaining, 'I have to be careful, for I dabble with poisons a good deal.' Could it be that Conan Doyle was drawing on another forensic scientist here, besides Christison? The chemical laboratory in the hospital, the blood analysis and dabbling with poisons bring to mind Alfred Swaine Taylor. St Bart's might be a simple substitution for Guy's. Besides, there are other aspects of Holmes that are strongly reminiscent of Taylor: the imposing height, the 'hawk-like nose', the sometimes difficult personality, the ability to reel off criminal cases at the drop of a hat.

In *The Adventure of the Speckled Band*, Holmes and Watson are investigating a doctor who is trying to poison his stepdaughter. Holmes says, 'When a doctor does go wrong he is the first of criminals. He has nerve and he has knowledge. Palmer and Pritchard were among the heads of their profession.' Palmer, of course, is The Rugeley Poisoner, and Pritchard was a doctor-poisoner who Christison helped to bring to justice. Conan Doyle was aware at least of Taylor's most famous case, but he mentions Taylor explicitly in his 1895 autobiographical novel *The Stark Monro Letters*.

Based on Conan Doyle's experience immediately after graduating from medical school, it features the eccentric James Cullingworth, based on Dr George Turnavine Budd. Cullingworth/Budd's practice was full of patients as he didn't charge for appointments; instead he fleeced them for medicine. In the novel, Cullingworth screams when one old lady enters his surgery:

'You've been drinking too much tea!' he cried. 'You are suffering from tea poisoning!'

Then, without allowing her to get a word in, he clutched her by her crackling black mantle, dragged her up to the table, and held out a copy of *Taylor's Medical Jurisprudence* which was lying there.

'Put your hand on the book,' he thundered, 'and swear that for fourteen days you will drink nothing but cocoa.'

Taylor's is not the only medical book to be mentioned in *The Stark Munro Letters*, although his is the only one treated as a sacred text. It's not likely that the average reader would recognise it; it was perhaps an in-joke.

That particular scene is set in 1882, the year Christison died, and the Holmes chronology sets the meeting of Holmes and Watson in the early 1880s. This is when Conan Doyle finished at Edinburgh Medical School, but it might be significant that at the same time as the Holmes stories began, two grand old men of medical jurisprudence died. Their obituaries were in newspapers and medical journals, and it doesn't seem an unlikely suggestion that their exploits fed into Conan Doyle's conception of his detective. Holmes's retort stands and bubbling test tubes are as much a part of the Holmes myth as the magnifying glass and deerstalker hat.

The obsolete Taylor

In 1907, twenty years after Holmes first appeared in print, Dr Thorndyke made his debut in R. Austin Freeman's *The Red Thumb Mark*. In the world of Freeman's fiction, Thorndyke is a lecturer of medical jurisprudence and toxicology, who qualified as a physician before becoming a barrister. Although Freeman initially claimed that Thorndyke was not based on anyone, he eventually admitted that the model for his scientific detective was Taylor, whose books he had read as a medical student. He even gave Thorndyke Taylor's height. Perhaps trying to throw the suspicious off his scent, Freeman damns his model with faint praise. Thorndyke has medical jurisprudence books on his shelves and advises Jervis, his Dr Watson, to read through them – 'Casper, Taylor, Guy, and Ferrier' – but Jervis objects.

'Casper and Taylor are pretty old, aren't they?'

'It is a capital mistake to neglect the authorities,' Thorndyke replies. 'Give your best attention to the venerable Casper and the obsolete Taylor and you will not be without your reward.'

When Thorndyke takes the witness stand at the end of *The Red Thumb Mark*, he does so very much in the style of Taylor, holding aloft for the court an enlarged image of the incriminating thumb mark. Taylor had taken test tubes and medical illustrations into the courtroom, visual aids important in explaining his analyses.

Freeman was obsessed with the idea of the scientific detective's importance to his stories, an early version of the forensic science-led novels and television series that we are so familiar with today. Thorndyke carries about with him a miniature portable laboratory: 'rows of little reagent bottles, tiny test-tubes, diminutive spirit-lamp, dwarf microscope and assorted instruments on the same Lilliputian scale'. It has all the tools he needs for the scientific detection of crime. Freeman's approach led to him inventing the inverted detective story, where the reader knows who the murderer is from the beginning, and suspense comes from watching Thorndyke employing his technologies, and his impressive brain, to solve the crime. For Freeman, the traditional tale 'The Singing Bone' had significance; it was forensic medicine foretold in folktale, the bone of a murder victim announcing the identity of the criminal.

Freeman believed that his readers were 'to be found among men of the definitely intellectual class', who would not be put off by Thorndyke's thorough analyses, or Freeman's prose, which author and critic Julian Symons described as 'chewing dry straw'. Freeman is somewhat obscure today; he was a eugenicist, and unfortunately his views intrude on his fiction. Lines such as 'The uncleanness of the criminal is not confined to his moral being; wherever he goes, he leaves a trail of actual, physical dirt' do not play well to a twenty-first-century audience, and neither do scenes in prisons where all the 'criminal class' prisoners are described as little more than apes, while the wrongly accused middle-class hero is presented as a more highly evolved being.

The back doors to death

After the bloody slaughter of The Great War, crime fiction evolved into the puzzle-dominated stories of The Golden Age. There was little blood and bludgeoning, and poisoning featured quite often as an unrealistically neat sort of death. What could Taylor, cutting up and boiling human intestines, have to do with these genteel crimes?

Dorothy L. Sayers' detective Lord Peter Wimsey briefly refers to 'a volume of Medical Jurisprudence' in *Whose Body?*, which sounds like one of Taylor's. In Sayers' final Wimsey mystery, the 1937 *Busman's Honeymoon*, she quotes at length from Taylor when a police officer is investigating a murder. On his bookshelves, 'tall and menacing, [there are] the two blue volumes of Taylor's *Medical Jurisprudence*, that canon of uncanonical practice and Baedeker of the back doors to death.' As the murder victim has head injuries, the police officer turns to the section on 'Intercranial Haemorrhage – Violence or Disease', and there follows a case from 1859, about a young man who fell out of a chaise.

Sayers inhabits the police officer for a moment, using the brief medical facts of the case to flesh out a whole existence for the 'excellent and unfortunate gentleman, his name unknown, his features a blank, his life a mystery; embalmed for ever in a fame outlasting the gilded monuments of princes!' Books on medical jurisprudence are still used as reference works for crime authors, who want to make sure their forensic science is correct, but the books also fire their imaginations.

The deceptive moon

One hundred and one years after the trial and execution of William Palmer, historical novelist Robert Graves gave the world *They Hanged My Saintly Billy*. Graves was charmed by the Rugeley Rogue, declaring him to be an innocent man condemned by an unfair trial. In order to convince readers of his thesis, Graves claimed that Palmer's wife committed suicide, and, rather like Palmer's defence team, went about undermining the prosecution.

Graves had at least read the strychnine section in the 1865 edition of *Principles and Practice*, but failed to understand that Taylor always compiled research from other scientists. 'None of this is the product of his own research – the Professor's light shines with borrowed rays, like the deceptive Moon.' But considering that Graves claims that the poisons register was checked for Palmer buying prussic acid, when only arsenic sales were recorded in 1855, and bearing in mind that throughout the novel he refers to his 'hero' as Dr Palmer, when he was a surgeon, not a physician, Graves's conclusions should be taken with a pinch of salt – or strychnine.

The ancestor

Alfred Swaine Taylor saw the opportunities presented by a new science and became its passionate, self-appointed champion, believing it promised the security of millions. Sometimes a controversial and outspoken figure, in the great majority of his cases he carefully weighed up all the evidence, helping to codify the way that the body and its secrets could solve crime, or reveal that no crime had been committed. Alfred Swaine Taylor is one of the ancestors of modern forensic science: he is part of its very DNA.

Timeline

1806	Alfred Swaine Taylor born in Northfleet, Kent on 6 December
1823	Enrols as a pupil at the United Hospitals of St Thomas's and Guy's in London
1831	Becomes professor of medical jurisprudence at Guy's Hospital
1832	Becomes joint professor of chemistry at Guy's, with Arthur Aikin; cholera epidemic
1833	Edwards inquest
1834	Marriage to Caroline Cancellor
1836	James Marsh's test for arsenic; volume 1 of Taylor's *Elements of Medical Jurisprudence*
1839	Begins photography experiments
1841	Hugo Reinsch's test for arsenic
1843	Reinsch's test begins to be used in Britain
1844	First edition of Taylor's *A Manual of Medical Jurisprudence* published.
1845–51	Editor of the *London Medical Gazette*
1845	Jennings and Tawell trials
1846	North trial
1847	First Chesham trial
1848	First edition of Taylor's *On Poisons in Relation to Medical Jurisprudence and Medicine*; May trial; Dore and Spry trial
1849	Geering trial; cholera epidemic
1851	Second Chesham trial; Drory trial; becomes sole lecturer of chemistry at Guy's
1852	Honorary MD from the University of St Andrew's
1853	Kirwan trial
1855	Wooler trial
1856	Palmer trial ('The Rugeley Poisoner')
1857	The Waterloo Bridge Mystery
1858	Arsenical wallpaper pigment danger revealed
1859	Second edition of *On Poisons*; awarded The Swiney Prize; Smethurst trial
1860	The Eastbourne Manslaughter
1862	Wilson trial

1863	Taylor and Brande's *Chemistry* published; research into the cooling of the body after death
1865	First edition of Taylor's *The Principles and Practice of Medical Jurisprudence*
1868	Baker trial ('Sweet Fanny Adams')
1870	Boulton and Park trial; Taylor retires as professor of chemistry and as an expert witness.
1873	Second edition of *The Principles and Practice of Medical Jurisprudence*
1875	Third edition of *On Poisons*
1876	Death of Taylor's wife
1878	Retires as Professor of Medical Jurisprudence and Toxicology
1879	Eighth edition of *A Manual of Medical Jurisprudence* published
1880	Dies in London on 27 May

Acknowledgements

Firstly, I would like to thank my editor, Linne Matthews, and everyone at Pen and Sword. I would also like to thank my partner, Gordon Wallace, who has put up with me regaling him each and every evening with Taylor's many delightful cases. Usually, while I cooked the dinner. Gordon also took many of the photographs in this book, and fuelled me with cups of tea (no arsenic discovered floating thereupon).

Many thanks to:
My friends and family (writers and otherwise) for their unflagging encouragement and interest in the face of many gruesome revelations.

Special Collections Division at the University of St Andrews; Main Library at the University of Edinburgh; Library Services and the Cadbury Research Library, Special Collections at the University of Birmingham; East Sussex Record Office; The British Library. The Wellcome Library, for digitising their collections and making them freely available online, along with rare books, journals and pamphlets belonging to the Royal College of Surgeons and the Royal College of Physicians in Edinburgh. The Hathi Trust, for trying to organise the mass of digitised historical books and journals which can be found online. Findmypast/British Newspaper Archive for making such a fund of historical newspapers available online. Guy's and St Thomas' NHS Foundation Trust for allowing me to include their portrait of Taylor in this book.

Alicia Zborowska for translating the titles of Taylor's two Italian articles into English. David Craze, retired Chemistry teacher, for his comments on Taylor's experiments in the Grotta del Cane. Balbir Rai Sharma, the helpful Northfleet cabby, and the friendly staff at Ebbsfleet International station. Fiona Orr, who told me that William Morris saw nothing wrong with arsenical wallpaper pigment. Rachel Brewster of Little Vintage Photography for the 'sun drawing'. Neil McKenna and Stephen Bates for corresponding with me. Reverend Anthony Perry of St Mary's, Bearwood, for lending me a copy of *The Book of Common Prayer*. Professor Ashlimann for permission to use an extract from his translation of 'The Singing Bone'. Luke at The Royal Parks for information on Regent's Park. And last but not least, Nick at Highgate Cemetery who rushed out into the rain to take a photo of Taylor's grave.

And of course, to you – my dear reader.

Further Reading

There are several in-depth books about cases that Taylor worked on. In chronological order:

My book *Poison Panic* covers the Essex arsenic trials of the 1840s.

The Poisoner by Stephen Bates will ably lead you through the life and crimes of William Palmer.

Ian Burney's *Poison, Detection, and the Victorian Imagination* explains the development of nineteenth-century toxicology and the arguments surrounding the Palmer and Smethurst trials.

Taylor makes a fleeting appearance in Kate Colquhoun's *Mr Briggs' Hat*, the story of Britain's first railway murder.

Neil McKenna's thought-provoking romp *Fanny & Stella: The Young Men Who Shocked Victorian England* is a must-read for anyone interested in hidden Victorian lives.

As a curiosity, Kate Summerscale's *The Wicked Boy* features one Charles Carne Lewis junior as a coroner; he is the son of Charles Carne Lewis senior, who frequently referred cases to Taylor.

Selected Bibliography

Alt, Laurence, 'Alfred Swaine Taylor (1801–80), Some Early Material', *History of Photography: an International Quarterly*, vol. 16, 1992, Winter, pp.397–8.

Anon, *Illustrated Life and Career of William Palmer, of Rugeley*, Ward & Lock, London, 1856.

Ballantine, William, *Some Experiences of a Barrister's Life*, Richard Bentley, London, 1883 (8th ed).

Barraud, H.R. & Jerrard, G.M.G., *The Medical Profession in All Countries*, vol. 1, J & A Churchill, London, 1873.

Barrell, Helen, *Poison Panic: Arsenic Deaths in 1840s Essex*, Pen & Sword, Barnsley, 2016.

Bates, Stephen, *The Poisoner: The Life and Crimes of Victorian England's Most Notorious Doctor*, Duckworth Overlook, London, 2015.

Besson, Alain, 'The Medico-Legal Tracts Collection of Dr A.S. Taylor, FRCP', *Journal of the Royal Society of Physicians of London*, vol. 17, April 1983, pp.147–9.

Borowitz, Albert, *The Bermondsey Horror: The Murder that Shocked Victorian Britain*, Robson Books, London, 1988.

Brande, William Thomas, & Taylor, Alfred Swaine, *Chemistry*, John W. Davies, London, 1863.

Browne, Lathom G. & Stewart, C.G., *Reports of Trials for Murder by Poisoning*, Stevens, London, 1883.

Burney, Ian, *Poison, Detection, and the Victorian Imagination*, Manchester University Press, 2012.

Cameron, H.C., *Mr Guy's Hospital: 1726–1948*, Longmans, London, 1954.

Christison, Robert, *A Treatise on Poisons, in Relation to Medical Jurisprudence, Physiology, and the Practice of Physic*, Adam Black, Edinburgh, 1832.

Christison, Robert, *A Treatise on Poisons in Relation to Medical Jurisprudence, Physiology, and the Practice of Physic*, Barrington & Haswell, Philadelphia, 1845.

Collins, Wilkie, *Armadale*, Smith, Elder, London, 1872.

Colquhon, Kate, *Mr Briggs' Hat: A Sensational Account of Britain's First Railway Murder*, Abacus, London, 2012.

Dickens, Charles, 'The Modern Alchemist', *All The Year Round*, 27 December 1862, pp.380–4.

Dudley-Edwards, Owen, *Burke and Hare*, Birlinn, Edinburgh, 2014.

Doyle, Arthur Conan, *The Stark Munro Letters*, Longmans, Green, London, 1895.

Doyle, Sir Arthur Conan, *Sherlock Holmes: The Complete Stories*, Wordsworth, London, 2007.

Emsley, Clive, *The English Police: a political and social history*, 2nd ed., Longman, London, 1996.

Flanders, Judith, *The Invention of Murder: How the Victorians Revelled in Death and Detection and Created Modern Crime*, Harper Press, London, 2011.

Forbes, Thomas Rogers, *Surgeons at the Bailey: English Forensic Medicine to 1878*, Yale University Press, London, 1985.

Freeman, R. Austin, *The Famous Cases of Dr Thorndyke*, Hodder & Stoughton, London, 1941.

Freeman, R. Austin, *The Red Thumb Mark. With a new introduction by Otto Penzler*, Mysteriouspress.com/Open Road, New York, 2014.

Golan, Tal, *Laws of Men and Laws of Nature: The History of Scientific Expert Testimony in England and America*, Harvard University Press, London, 2004.

Graves, Robert, *They Hanged My Saintly Billy*, Xanadu, London, 1989.

Knott, George, H., revised by Watson, Eric R., *Trial of William Palmer*, William Hodge, London, 1952.

Mant, A. Keith, ed., *Taylor's Principles and Practice of Medical Jurisprudence*, Churchill Livingstone, Edinburgh, 1984.

Marsh, James, 'Account of a method of separating small quantities of arsenic from substances with which it may be mixed', *Edinburgh New Philosophical Journal*, 1836, vol. 21, pp.229–36.

McKenna, Neil, *Fanny & Stella: The Young Men Who Shocked Victorian England*, Faber & Faber, London, 2014.

Parry, Leonard E., ed., *Trial of Dr Smethurst*, William Hodge & Co., Edinburgh & London, 1931.

Pearsall, Ronald, *Conan Doyle: A Biographical Solution*, Weidenfeld & Nicolson, London, 1977.

Priestman, Martin, ed., *The Cambridge Companion to Crime Fiction*, Cambridge University Press, 2003.

Rewcastle, James, *A Record of the Great Fire of Newcastle and Gateshead*, George Routledge, London, 1855.

Robertson, William Tindal, ed., *Photographs of Eminent Medical Men of All Countries, with Brief Analytical Notices of Their Works*, vol. 1, John Churchill, London, 1868.

Sayers, Dorothy L., *Whose Body?*, Boni & Liveright, New York, 1923.

Sayers, Dorothy L., *Busman's Honeymoon*, Victor Gollancz, London, 1937.

Schaaf, Larry J., *Records of the Dawn of Photography: Talbot's Notebooks P and Q*, Cambridge University Press, 1996.

Smith, Sydney, ed., *Taylor's Principles and Practice of Medical Jurisprudence*, 10th ed., Churchill, London, 1948.

Stevenson, Thomas, ed., *Principles and Practice of Medical Jurisprudence*, Henry Lea, Philadelphia, 1883.

Symons, Julian, *Bloody Murder: From the Detective Story to the Crime Novel: A History*, Penguin, Harmondsworth, 1974.

Taylor, Alfred Swaine, 'Poisoning by sulphuric acid', *The London Medical and Physical Journal*, New Series 73, July 1832, pp.1–4.

Taylor, Alfred Swaine, 'Case, illustrative of the Contagious Propagation of Cholera', *The London Medical and Physical Journal*, New Series 73, August 1832, pp.155–19.

Taylor, Alfred Swaine, 'Remarks on the Law of England as it relates to the Crime of Foeticide', *The London Medical and Physical Journal*, New Series 73, September 1832, pp.184–98.

Taylor, Alfred Swaine, 'An Account of the Grotta del Cane; with Remarks on Suffocation by Carbonic Acid', *The London Medical and Physical Journal*, New Series 73, October 1832, pp.278–85.

Taylor, Alfred Swaine, 'On the Formation of Sulphur at the Solfatara, near Naples', *The London Medical and Physical Journal*, New Series 73, October 1832, pp.343–4.

Taylor, Alfred Swaine, *Elements of Medical Jurisprudence*, Deacon, London, 1836.

Taylor, Alfred Swaine, 'Two cases of fatal poisoning by arsenious acid: with remarks on the solubility of that poison in water and other menstrua', *Guy's Hospital Reports*, vol. 2, 1837, pp.68–103.

Taylor, Alfred Swaine, *On the Art of Photogenic Drawing*, Jeffery, London, 1840.

Taylor, Alfred Swaine, 'On poisoning by arsenic: the quantity required to destroy life', *Guy's Hospital Reports*, 1841, series 1, vol. vi, pp.21–38.

Taylor, Alfred Swaine, 'Medico-legal report of the evidence given on a recent trial for murder by poisoning with arsenic', *Guy's Hospital Reports*, 1841, series 1, vol. vi, pp.265–96.

Taylor, Alfred S., 'Report on the New Test for Arsenic, and its Value compared with the other Methods of detecting that Poison', *The British & Foreign Medical Review or Quarterly Journal of Practical Medicine & Surgery*, vol. 15, July 1843, pp.275–82.

Taylor, Alfred Swaine, *Elements of Medical Jurisprudence, Interspersed with a Copious Selection of Curious Cases and Analyses of Opinions Delivered at Coroners' Inquests*, Deacon, London, 1843.

Taylor, Alfred Swaine, *A Manual of Medical Jurisprudence*, John Churchill, London, 1844.

Taylor, Alfred Swaine, *A Thermometrical Table on the Scales of Fahrenheir, Centrigrade, and Réaumur*, Thomas & Richard Willats, London, 1845.

Taylor, Alfred Swaine, 'Cases in Medical Jurisprudence, with remarks', *Guy's Hospital Reports*, series 2, vol. 3, 1845, pp.39–75.

Taylor, Alfred Swaine, 'Trial for murder by poisoning with arsenic, Berkshire Lent Assizes, 1845', *Guy's Hospital Reports*, 1845, series 2, vol. 3, pp.187–96.

Taylor, Alfred Swaine, *On the temperature of the earth and sea, in reference to the theory of central heat*, Wilson & Ogilvy, London, 1846.

Taylor, Alfred Swaine, 'Cases and observations in medical jurisprudence,' *Guy's Hospital Reports*, October 1846, series 2, vol. 4, pp.396–489.

Taylor, Alfred Swaine, *On Poisons in Relation to Medical Jurisprudence and Medicine*, Blanchard & Lea, Philadelphia, 1848.

Taylor, Alfred Swaine, *Medical Jurisprudence*, ed. Hartshorne, Edward, Blanchard & Lea, Philadelphia, 1853, 3rd US ed. from 4th British ed. of *Manual of Medical Jurisprudence*.

Taylor, Alfred Swaine, 'Analysis of the water of the Great Geyser, Iceland', *Guy's Hospital Reports*, series 3, vol. 2, 1856, pp.405–407.

Taylor, Alfred Swaine, *On Poisoning by Strychnia, with Comments on the Medical Evidence Given at the Trial of William Palmer for the Murder of John Parsons Cook*, Longman, Brown, Green, Longmans & Roberts, London, 1856.

Taylor, Alfred Swaine, 'On poisoning by tartarized antimony', *Guy's Hospital Reports*, series 3 vol. 3, 1857, pp.369–481.

Taylor, Alfred Swaine, 'On the detection of absorbed strychnia and other poisons', *Guy's Hospital Reports*, series 3, vol. 3, 1857, pp.482–501.

Taylor, Alfred Swaine, *On Poisons in Relation to Medical Jurisprudence and Medicine*, Blanchard & Lea, Philadelphia, 1859.

Taylor, Alfred Swaine, 'Ophthalmia as a result of the use of arsenical wallpapers', *Ophthalmic Hospital Reports & Journal of the London Ophthalmic Hospital*, no. 6, January 1859.

Taylor, Alfred Swaine, 'Facts and fallacies connected with the research for arsenic and antimony', *Guy's Hospital Reports*, series 3, vol. 6, 1860, pp.201–65.

Taylor, Alfred Swaine, *The Manual of Medical Jurisprudence*, John Churchill, London, 1861.

Taylor, Alfred Swaine & Wilks, Samuel, 'On the cooling of the human body after death: inferences respecting the time of death. Observations of temperature made in 100 cases', *Guy's Hospital Reports*, series 3, vol. 9, 1863, pp.180–211.

Taylor, Alfred Swaine, *The Principles and Practice of Medical Jurisprudence*, John Churchill, London, 1865.

Taylor, Alfred Swaine, 'On Homicidal and Suicidal Wounds of the Throat', *Guy's Hospital Reports*, series 3, vol. 14, 1869, pp.112–44.

Taylor, Alfred Swaine, *Metropolis Water Supply. Remarks on the water supplied by the West Middlesex Water Works Company*, London, 1872.

Taylor, Alfred Swaine, *The Principles and Practice of Medical Jurisprudence*, 2nd US ed., Henry C. Lea, Philadelphia, 1873.

Taylor, Alfred Swaine, 'Medico-legal observations on tattoo-marks as evidence of personal identity. Remarks on the Tichborne case', *Guy's Hospital Reports*, series 3, vol. 19, 1874, pp.441–65.

Taylor, Alfred Swaine, 'Death from disease or poison: Does the retention or maintenance of heat in a dead body furnish any indication of the cause of death?', *Guy's Hospital Reports*, series 3, 19, 1874, pp.468–87.

Taylor, Alfred Swaine, 'On the detection of blood by guaiacum', *Guy's Hospital Reports*, series 3, vol. 19, 1874, pp.517–19.

Taylor, Roger, *Impressed by Light: British photographs from paper negatives, 1840–1860. Biographical dictionary by Larry J. Schaaf in collaboration with Roger Taylor*, Metropolitan Museum of Art, New York, 2007.

Thomson, Anthony Todd & Amos, Andrew, *Syllabus of Lectures on Medical Jurisprudence in the University of London*, Joseph Mallett, London, 1830.

Watson, Katherine, *Poisoned Lives: English Poisoners and their Victims*, Hambledon & London, London, 2004.

White, Stephen, 'Alfred Swaine Taylor – A Little Known Photographic Pioneer', *History of Photography: an International Quarterly*, vol. 11, 1987, July–Sept, pp.229–35.

Whorton, James C., *The Arsenic Century: How Victorian Britain was Poisoned at Home, Work, and Play*, Oxford University Press, 2010.

Wilks, Samuel & Bettany, G.T., *A Biographical History of Guy's Hospital*, Ward, Lock, Bowden & Co., London, 1892.

Notes

Genealogy and census information compiled from Findmypast, Ancestry, FreeBMD, FreeREG, www.cityark.medway.gov.uk, East Sussex Record Office.

Extract from 'The Singing Bone', collected by The Brothers Grimm, translated by Professor D.L. Ashliman http://www.pitt.edu/~dash/grimm028.html. Used with permission.

Introduction

The boots: *Cambridge Independent Press*, 1 August 1863, Taylor, A.S., *Principles and Practice of Medical Jurisprudence*, John Churchill, London, 1865, p.429. **History of medical jurisprudence:** Knight, B., 'The development of medico-legal systems' in Mant, A Keith, ed, *Taylor's Principles and Practice of Medical Jurisprudence*, Edinburgh & London: Churchill Livingstone, 1984, pp.11–13.

Chapter 1

Taylor's early years: Earles, M.P., 'Taylor, Alfred Swaine (1806–1880)', *Oxford Dictionary of National Biography (ODNB)*, Oxford University Press, 2004; Obituary: Alfred Swaine Taylor, MD, FRS, *The British Medical Journal*, 12 June 1880, pp.905–906; Barraud, H.R. & Jerrard, G. M.G., *The Medical Profession in All Countries*, vol. 1., J & A Churchill, London, 1873, unpaginated; Wilks, Samuel & Bettany, G.T., *A Biographical History of Guy's Hospital*, Ward, Lock, Bowden & Co., London, 1892, pp.392–5. **Northfleet:** Population and East India Company's presence: www. discovergravesham.co.uk; Watercress: *The Pharmaceutical Journal*, 14 June 2008, vol. 280, p. 727; *The Saturday Magazine*, no. 435, 13 April 1839. Many biographies of Taylor say that his father was a captain in the East India Company from King's Lynn, but he does not appear in Farrington's *A Biographical Index of East India Company Maritime Service Officers: 1600–1834*. However, the inscription on his headstone says 'late of the Hon. E.I. Company's Maritime Service'. **Marriage of Alfred Swaine Taylor's parents** at St Botolph's, Northfleet, on 15 September 1804, Thomas Taylor & Susannah Bacheldor Badger (Cityark). Taylor's mother signed her name 'Sukey Bacheldor Badger' at her marriage, as she had also been baptised. Her father's will (Charles Badger, PCC PROB 11/1341/249) named her as 'Susannah Batchelor Badger'. Her headstone reads just 'Susan'. Headstone inscription: www.kentarchaeology. org.uk. Thomas Taylor's second marriage probably took place in 1820 at St James, Clerkenwell, to Elizabeth Knight, a widow. **Oxley and Taylor:** advert in *Public Ledger & Daily Advertiser*, 27 October 1818. **Albemarle House school:** Adverts: *Windsor & Eton Express*, 30 Dec 1837. *Morning Advertiser* 6 August 1842. Bensly, Edward, 'Albemarle House, Hounslow', *Notes and Queries*, 6 March 1937. Walford, Edward,

Greater London: A Narrative of Its History, Its People, and Its Places, vol. 1., Cassell, London, 1883, pp.60–9. Anon, 'Familiar Verses, Addresses to Two Young Gentlemen at the Hounslow Academy', *An Asylum for Fugitive Pieces, in Prose and Verse, Not in Any Other Collection, With Several Pieces Never Before Published*, Debrett, London, 1785, p.86. **Apprenticeship and medical training:** Thames Watermen & Lightermen 1688–2010, Binding Records, Silas Badger Taylor bound to Thomas Elkin, 6 Jan 1825; Cameron, H.C., *Mr Guy's Hospital: 1726–1948*, Longmans, London, 1954, pp.9–10, pp.80–5, 95; Copeman, W.S.C., *The Worshipful Society of Apothecaries of London: A History 1617–1967*, Pergamon, London, 1967, p.56, 67. Genealogical sources show that Taylor's master was Donald Macrae, originally from Scotland. **Taylor's early life and his publication history:** Robertson, William Tindal, ed., *Photographs of Eminent Medical Men of All Countries, with Brief Analytical Notices of Their Works*, vol. 2, John Churchill, London, 1868, pp.37–41. **Continental travel:** Towner, John, *The European Grand Tour*, Doctoral (PhD) thesis, University of Birmingham, 1985, pp.341–3, 351. Taylor, Alfred Swaine, 'On the Formation of Sulphur at the Solfatara, near Naples', *The London Medical and Physical Journal*, New Series 73, October 1832, pp.343–4; Taylor, Alfred Swaine, 'An Account of the Grotta del Cane; with Remarks on Suffocation by Carbonic Acid', *The London Medical and Physical Journal*, New Series 73, October 1832, pp.278–85. **History of Guy's and St Thomas's:** Cameron, ibid., p.58, 79, 90, 92. Allen & Aikin, p.286. **Teaching of medical jurisprudence:** Beck: Taylor, Alfred Swaine, 'Preface to the American edition', *The Principles and Practice of Medical Jurisprudence*, vol. 1, Henry C. Lea, Philadelphia, 1873, p.i. Burney, Ian, *Poison, detection, and the Victorian imagination*, Manchester University Press, 2012, pp.40–3. Forbes, Thomas Rogers, *Surgeons at the Bailey: English Forensic Medicine to 1878*, Yale University Press, London, 1985, pp.4–5, 7–9. Christison: Burke & Hare trial: Dudley-Edwards, Owen Dudley, *Burke and Hare*, Birlinn, Edinburgh, 2014, p.158; career: White, Brenda M., 'Christison, Sir Robert, first baronet (1797–1882)', *ODNB*. Smith: Ward, Ward, 'Smith, John Gordon (1792–1833)', *ODNB*; Smith, John Gordon, 'Introductory lecture on medical jurisprudence', *The Lancet*, 16 October 1830, pp.97–103. Adverts for medical jurisprudence lectures: *Durham County Advertiser*, 26 August 1831; *Morning Post*, 16 August 1831, ibid., 29 August 1831. **Taylor's appearance:** Cameron, ibid., p.386; portrait c. 1840 at Guy's Hospital. **Thomson and Amos:** Thomson, Anthony Todd & Amos, Andrew, *Syllabus of Lectures on Medical Jurisprudence in the University of London*, Joseph Mallett, London, 1830; *The Lancet*, 12 February p.561, 19 February 1831, pp.690–694.

Chapter 2

Taylor's medical practice, publishing in the *LM&PJ:* Robertson, ibid. Address on Great Marlborough Street: *The Times*, 25 January 1833. **Cases:** Russell: Moody, William, 'Rex v Henry Russell', *Crown Cases Reserved for Consideration; and Decided by the Judges of England from the year 1824, to the year 1837*, vol. 1, Johnson, Philadelphia, 1839. pp.356–68; *Morning Herald* quoted in the *London Evening Standard*, 5 May 1832, p.4. Kinnear case: *The Lancet*, 14 & 21 July 1832. Isabella Creswell: *Morning Chronicle* and *Bell's London Life and Sporting Chronicle*, 10 June 1832. Taylor, Alfred Swaine, 'Poisoning by Sulphuric Acid, *LMPJ*, NS 73, July 1832, pp.1–4. The John

Snow Site (http://www.ph.ucla.edu/epi/snow/testimonials.html). French was the medical officer of St James's Infirmary from 1830–72. **Taylor's *LMPJ* articles:** The 1832 volume gives us a sample of his output and the wide range of subjects he covered. Above, and: 'Case, illustrative of the Contagious Propagation of Cholera', August 1832, pp.155–9; 'Remarks on the Law of England as it relates to the Crime of Foeticide', September 1832, pp.184–98. **Suicide of medical student:** *Bell's Life in London and Sporting Chronicle*, 16 January 1831. **Alexander Barry:** *The Spectator*, 8 September 1832, p.7. 'Proceedings of Learned Societies': Royal Society. Address of his Royal Highness the President, delivered at the Anniversary Meeting, 10 November, 1832', XXIII *The London and Edinburgh Philosophical Magazine and Journal of Science*, vol. 1, p.140. **Arthur Aikin:** Torrens, H.S., 'Aikin, Arthur (1773–1854)', *ODNB*; Cameron, ibid., p.286, says Aikin began lecturing at Guy's in 1826, but Torrens says it was 1821. Aikin appears in Allen's diary in 1822 so Torrens is probably correct. [*Life of William Allen with Selections from his Correspondence in Two Volumes*, vol. 2, Henry Longstreth, Philadelphia, 1847, p.41] Allen's entry for 15 August 1826 says that Barry has replaced him (ibid., p.170). **Eliza Edwards:** *Morning Chronicle*, 25 January; *The Times*, 25 January 1833; Rictor Norton (ed.), 'The Extraordinary Story of "Eliza" Edwards, 1833', *Homosexuality in Nineteenth-Century England: A Sourcebook*, 27 May 2012 (http://rictornorton.co.uk/eighteen/edwards.htm); Taylor, Alfred Swaine, *A Manual of Medical Jurisprudence*, John Churchill, London, 1861, p.657. James Barry: Hurwitz, Brian & Richardson, Ruth, 'Inspector General James Barry MD: Putting the Woman in Her Place', *BMJ*, 4 February 1989, pp.299–305.

Chapter 3
Marriage: 14 July 1834, *Public Ledger & Daily Advertiser;* parish register (London Metropolitan Archives). Domestic tastes: *BMJ*, 12 June 1880, p.906. **Caroline Cancellor's family:** Richard Hall: Walford, Edward, *The County Families of the United Kingdom*, Robert Hardwicke, London, 1860, p.103. Her mother's death: *Royal Cornwall Gazette*, 17 March 1810. John Cancellor: memorial inscription: 'St Pancras Church', in *Survey of London: Volume 24, the Parish of St Pancras Part 4: King's Cross Neighbourhood*, ed. Walter H. Godfrey & W. McB. Marcham (London, 1952), pp.1–9; will PCC PROB 11/1794/114; donations: *The Times*, 9 October 1798; 20 May 1822. John Henry Cancellor: He was 'of Gray's Inn' when he married in 1829. He became a Master of the Court of Common Pleas in 1837. Richard Cancellor: will PCC PROB 11/1826/90. Burial record: Sussex Family History Group, via Findmypast. **Maps of the Regent's Park area:** Holmes, Malcolm, *Old Ordnance Survey Maps: The Godfrey Edition, Euston & Regent's Park 1870, London Sheet 49*, Leadgate, Consett, 2010. *Kelly's Post Office Directory Map of London* 1857. **Richard Alfred Taylor:** St Pancras baptism register, St James burial register. **Reviews of *Elements of Medical Jurisprudence*:** *London Medical and Surgical Journal*, 26 March 1836, pp.275–6; *The Brighton Gazette*, 12 May 1836. **Visit to Brighton:** *Morning Advertiser*, 2 August 1836. **The Swinleys and Atkinsons:** Marriage of Silas Badger Taylor and Mary Ann Swinley: ESRO, PAR 255/1/3/9 entry number 731. Marriage of William Atkinson and Louisa Swinley: ESRO, PAR 255/1/3/7 entry number 584. William Atkinson's chemical company: http://catalyst.org.uk/ (Science Discovery Centre, Widnes) 'Part

One 1654–1884: Primatt to Atkinson'. *Guy's Hospital Reports (GHR)*: Cameron, ibid., p.109, 129, 231. Taylor's entry in the index, 1836–93, *GHR*, vol. 50 (series 3, vol. 35), pp.723–5. **The Marsh test:** Christison, Robert, *A Treatise on Poisons, in Relation to Medical Jurisprudence, Physiology, and the Practice of Physic*, Adam Black, Edinburgh, & Longman, Rees, Orme, Brown & Green, London, 1832, pp.223–53. James, Frank A.J.L., 'Marsh, James (1794–1846)', *ODNB*. Marsh, James, 'Account of a method of separating small quantities of arsenic from substances with which it may be mixed', *Edinburgh New Philosophical Journal*, 1836, vol. 21, pp.229–36. Christison, Robert, *A Treatise on Poisons in Relation to Medical Jurisprudence, Physiology, and the Practice of Physic*, Barrington & Haswell, Philadelphia, 1845, pp.211–14. Taylor, Alfred Swaine, 'Two cases of fatal poisoning by arsenious acid: with remarks on the solubility of that poison in water and other menstrua', *GHR*, vol. 2, 1837, pp.68–103. Identity of Ellen Dickson: *Bell's New Weekly Messenger*, 11 September 1836. Review of Taylor's article: *British and Foreign Medical Review*, October 1837, pp.358–64. **Taylor at King's College:** Advert in *The Times*, 17 May 1838.

Chapter 4
Early photography: Schaaf, Larry J., 'Talbot, William Henry Fox (1800–1877)', *ODNB.*; Talbot, H.F. (Henry William Fox), 'Some account of the Art of Photogenic Drawing, or the Process by which Natural Objects may be made to delineate themselves without the aid of the artist's pencil'. *Abstracts of the Papers Printed in the Philosophical Transactions of the Royal Society of London*, vol. 4 (1837–1843), pp.120–1; White, Stephen, 'Alfred Swaine Taylor – A Little Known Photographic Pioneer', *History of Photography: an International Quarterly*, vol. 11, 1987, July-September, pp.229–35; Taylor, Alfred Swaine, *On the Art of Photogenic Drawing*, Jeffery, London, 1840. Alt, Laurence, 'Alfred Swaine Taylor (1801–80), Some Early Material', *History of Photography: an International Quarterly*, vol. 16, Winter 1992, pp.397–8 [year of birth incorrect in article title]. *Impressed by Light: British photographs from paper negatives, 1840–1860*, by Roger Taylor with a biographical dictionary by Larry J. Schaaf in collaboration with Roger Taylor, Metropolitan Museum of Art, New York, 2007, p.378. Schaaf, Larry J., *Records of the Dawn of Photography: Talbot's Notebooks P and Q*, Cambridge University Press, 1996, unpaginated. 'Our Weekly Gossip', *The Athenaeum*, no. 670, 29 August 1840, p.684. Taylor's photographs and photogenic drawings cannot be reproduced as I have been unable to contact albums' owners. Alt suggests this was the back of houses on Albany Street, but the gardens in the photograph seem too long. **Thomas Taylor:** death announcement: *London Evening Standard*, 30 September 1840; burial record (Cityark). Will PCC PROB 11/1937/19. Basil was presumably named after Alfred and Silas's great-grandfather, Basil Hunt (1728–1801), a carpenter from Lenham, Kent. **Oxley and Taylor:** as gun traders: advert in Cornish, James, *Cornish's Grand Junction and the Liverpool and Manchester Railway Companion*, 1837, Birmingham & London, p.146; as travel agents: *The Suffolk Chronicle*, 13 May 1837. **Arsenic in the wine:** Taylor, Alfred Swaine, 'On poisoning by arsenic: the quantity required to destroy life', *GHR*, 1841, series 1, vol. vi, pp.21–38. **Rhymes case:** *Evening Mail*, 1 March 1841. Taylor, Alfred Swaine, 'Medico-legal report of the evidence given on a recent trial for murder by poisoning with arsenic',

GHR, 1841, series 1, vol. vi, pp.265–96. County Constabulary Acts: Clive Emsley, *The English Police: a political and social history*, 2nd ed., Longman, London, 1996. **Marie-Fortunée Lafarge:** Taylor, Alfred Swaine, *On Poisons in Relation to Medical Jurisprudence and Medicine*, Blanchard & Lea, Philadelphia, 1859, p.374.

Chapter 5

A Manual of Medical Jurisprudence: The Lancet correspondence: 8 October 1842, p.80; 15 October 1842, p.839. Bartrip, P.W.J., 'Churchill, John Spriggs Morss (1801–1875)', *ODNB*. Reviews of: *The Lancet*, 27 April 1844, pp.159–61; *Provincial Medical and Surgical Journal*, 6 January 1844, pp.271–3; *The London Medical Gazette*, 15 November 1844, p.227. Writing process: Besson, Alain, 'The Medico-Legal Tracts Collection of Dr A.S. Taylor, FRCP', *Journal of the Royal Society of Physicians of London*, vol. 17, April 1983, pp.147–9. Obituary, Alfred Swaine Taylor, MD, FRS, *BMJ*, 12 June 1880, pp.905–906. **The Reinsch test:** Reinsch, Hugo, 'Ueber das Verhalten des metallischen Kupfers zu einigen Metalllösungen', *Journal für Praktische Chemie, 1841*, vol. 24 (1), pp.244–50; Taylor, Alfred S., 'Report on the New Test for Arsenic, and its Value compared with the other Methods of detecting that Poison', *The British & Foreign Medical Review or Quarterly Journal of Practical Medicine & Surgery*, vol. 15, July 1843, pp.275–82; Christison, Robert, 'Dr Christison on the New Mode of Detecting Arsenic', *The Lancet*, 16 September 1843, pp.870–1 [from *London and Edinburgh Journal of Medical Science*, September 1843]. Doubt in poisoning trials following Orfila's claim: Taylor, Alfred Swaine, *On Poisons...*, Blanchard & Lea, Philadelphia, 1859, pp.371–2. **Opium deaths:** Vaughan: *The Morning Post*, 25 & 29 March 1843; Ford: *Morning Chronicle*, 1 & 7 December 1843. **Birth of Edith Caroline Taylor:** St Pancras, baptism register. **The Leaver/Lever case:** *Old Bailey Online*; Taylor, Alfred Swaine, *On Poisons* (ibid.), p.347, 363. **Dr Guy:** His letter to *The Lancet*, 10 October 1844, p.117. Guy's defence: *[The London] Medical Review*, 29 November 1844, pp.298–302. Taylor's letter: *[The London] Medical Review*, 6 December 1844, p.335. Reviews of *Principles of Forensic Medicine*: *The London Medical Gazette*, 15 November 1844, pp.227–9; *The Lancet*, 15 February 1845, pp.188–9. Biography: Bettany, G.T., 'Guy, William Augustus (*bap.* 1810, *d.* 1885)', rev. Richard Hankins, *ODNB*. **London Medical Gazette:** Taylor's obituary in the *British Medical Journal* (ibid.) says he was editor from 1845–51. Sample of the *London Medical Gazette* under Taylor's editorship from issues published July 1846 to February 1847. Act for the Attendance and Remuneration of Medical Witnesses at Coroners' Inquests (1836): Watson, Katherine, *Poisoned Lives: English Poisoners and their Victims*, Hambledon & London, London, 2004, pp.165–6.

Chapter 6

Thermometer: Taylor, Alfred Swaine, *A Thermometrical Table on the Scales of Fahrenheit, Centigrade, and Réaumur*, Thomas & Richard Willats, London, 1845; *London Medical Gazette*, 31 January 1845, pp.587–8. **Jennings case:** Taylor, Alfred Swaine, 'Trial for murder by poisoning with arsenic, Berkshire Lent Assizes, 1845', *GHR*, 1845, (vol. 10), series 2, vol. 3, pp.187–96; *Wiltshire Independent*, 13 February 1845; *Reading Mercury*, 22 March 1845. **Tawell case:** *The Times*, 4, 6 8, 15, 16, 31

January; 6, 13, 14, 15, 17, 20, 22, 29 March 1845. 'God's Lightning' – from *Punch*, reporting on the Bermondsey Horror in 1849, which also used the telegraph to track down suspects (quoted in *Glasgow Herald*, 7 September 1849); 'John Tawell: The Man Hanged By the Electric Telegraph', University of Salford, http://www.cntr.salford. ac.uk/comms/johntawell.php; James, Frank A.J.L., 'Cooper, John Thomas (1790–1854)', *ODNB*. Prussic acid: from Taylor's, *On Poisons*, 1859, ibid., p.562. Smell: pp.567–8 and genetics, Online Mendelian Inheritance of Man; Taylor, Alfred Swaine, 'Cases in Medical Jurisprudence, with remarks', *GHR*, series 2, vol. 3, 1845, pp.39–75. Letheby, Henry, 'Remarks on poisoning by prussic acid', *The Lancet*, 3 May 1845, pp.497–8. Allen, C.J.W., 'Kelly, Sir Fitzroy Edward (1796–1880)', *ODNB*. Hamlin, Christopher, 'Letheby, Henry (1816–1876)', *ODNB*. **Hobson:** *The Ipswich Journal*, 30 March 1844; Taylor's article in *GHR*, series 2, vol. 3, pp.39–75. **Christison:** White, Brenda, ibid. His addresses: He appears at 3 Great Stuart Street in *The Post Office Annual Directory 1832–33*, Edinburgh, 1832, p.36; at 5 Randolph Crescent on the 1841 census, and at 40 Moray Place from the 1851 census until his death (*Edinburgh Evening News*, 27 January 1887). **Ellen Fry:** *Reading Mercury*, 2 August 1845; Taylor, Alfred Swaine, 'Case of poisoning by arsenic: Cases and observations in medical jurisprudence', *GHR*, October 1846, series 2, vol. 4, pp.458–64. **West Bromwich case:** Taylor, Alfred Swaine, 'Case of poisoning by the essential oil of bitter almonds: Cases and observations in medical jurisprudence', *GHR*, October 1846, series 2, vol. 4, pp.478–89.

Chapter 7

Taylor's lecture: Taylor, Alfred Swaine, *On the temperature of the earth and sea, in reference to the theory of central heat*, Wilson & Ogilvy, London, 1846. **North case:** Taylor, Alfred Swaine, 'Report of a trial for murder by poisoning with oil of vitriol, with remarks: Cases and observations in medical jurisprudence', *GHR*, October 1846, series 2, vol. 4, pp.396–443; *The Monmouthshire Merlin*, 1 August 1846; *The Lancet*, 12 September 1846, p.304; 26 September 1846, p.359. 'Nutrient enemas: an example of one being used in a fever case': *London Medical Gazette*, 27 April 1833, pp.106–10. **Chesham cases:** Barrell, Helen, *Poison Panic: Arsenic Deaths in 1840s Essex*, Pen & Sword, Barnsley, 2016, pp.1–46. **Poisonings:** Taylor, Alfred Swaine, 'Cases and observations in medical jurisprudence', *GHR*, October 1846, series 2, vol. 4, pp.396–492. **Secret poisonings:** *London Medical Gazette*, editorials: 15 January 1847, pp.105–108; 29 January 1847, pp.191–4; Review of Bulwer-Lytton's *A Word to the Public*, 5 February 1847, pp.242–8. Poisoned plum cake: *Aris's Birmingham Gazette*, 23 November 1840.

Chapter 8

Bowyer: *Suffolk Chronicle*, 13 May 1848; *Essex Standard*, 12 May 1848; *Bury & Norwich Post* 9 August 1848; *London Evening Standard* 4 & 5 August 1848; *Chelmsford Chronicle*, 11 August 1848. **May:** Barrell, ibid., pp.47–83. Friendship of Pollock and Taylor: Parry, Leonard E., ed., *Trial of Dr Smethurst*, William Hodge & Co, Edinburgh & London, 1931, p.162, & Hannavy, John, *Encyclopedia of Nineteenth-Century Photography, vol. 2*, Routledge, Abingdon, 2008. p.1144. **Dore and Spry:**

Morning Advertiser, 29 July 1848; *Morning Chronicle*, 2 August 1848; *Lloyd's Weekly Newspaper*, 6 August 1848; *Chelmsford Chronicle*, 11 August 1848; *Bristol Mercury*, 12 August 1848; *London Evening Standard*, 16 August 1848; *London Evening Standard*, 22 August 1848; *Morning Advertiser*, 22 March 1852; *Old Bailey Online*. Mary Ann Dore's husband, George Warrener Dore, was born on the Isle of Wight, (1851 census), baptised at Carisbrooke (1847). Intussusception: Great Ormond Street Hospital for Children, www.gosh.nhs.uk. **Southgate and Button**: Barrell, ibid., pp.84–127. **On Poisons**: Taylor, Alfred Swaine, *On Poisons in Relation to Medical Jurisprudence and Medicine*, Blanchard & Lea, Philadelphia, 1848, p.v.

Chapter 9
John Snow: Vinten-Johansen et al, *Cholera, Chloroform, and the Science of Medicine: A Life of John Snow*, Oxford University Press, 2003, p.109 (note), and pp.140–64. Boy's death: Taylor, *On Poisons*, ibid., 1859, p.654. **Southgate**: Barrell, ibid., pp.84–127; Ballantine, William, *Some Experiences of a Barrister's Life*, Richard Bentley, London, 1883 (8th ed), p.109. **Geering**: Watson, ibid., pp.87–8; *London Evening Standard*, 15 May 1849; *Chelmsford Chronicle*, 18 May 1849; *Shrewsbury Chronicle*, 10 August 1849. **Mary Cancellor**: PCC will: PROB 11/2097/224. **Cholera**: *GHR*, vol. ix, July–December 1849. **The Mannings**: Boase, G.C., 'Manning, Marie (1821-1849)', rev. J. Gilliland, *ODNB*. [She was known as Maria to the British, but born Marie.]. Russell, Colin A., 'Odling, William (1829–1921)', rev. *ODNB*. Borowitz, Albert, *The Bermondsey Horror: The Murder that Shocked Victorian Britain*, Robson Books, 1988, pp.101–102. *The Morning Post*, 25 August 1849; *Sussex Advertiser, Surrey Gazette*, 28 August 1849. Ballantine, ibid., p.156. **Lucas and Reeder**: *Cambridge Independent Press*, 30 March 1850. *Bell's New Weekly Messenger*, 31 March 1850. Taylor, *On Poisons*, 1859, ibid., p.380. Probert, Rebecca, *Marriage Law for Genealogists: The Definitive Guide*, Takeaway, Kenilworth, 2012, p.66. **Hartley**: Taylor, *On Poisons*, 1859, ibid., pp.188–9; *Old Bailey Online*; *Morning Post*, 13 May 1850. **Chesham**: Barrell, ibid., pp.128–61; *Chelmsford Chronicle*, 24 May 1850. **Bristow**: *Lincolnshire Chronicle*, 25 October 1850.

Chapter 10
Drory: *The Times*, 15, 16 & 18 October, 4 November 1850; *Chelmsford Chronicle*, 25 October 1850; Barrell, ibid. **Household Words**: Fitzgerald, Percy Hetherington, 'Fleur de Lys', *Household Words*, 23 October 1850, p.441. From *Dickens Journals Online*. **Wren**: *The Hampshire Advertiser*, 4 January 1851; *Hampshire Telegraph and Sussex Chronicle*, 8 March. Taylor, *On Poisons*, 1859, ibid., p.673. **Chesham and Drory, Sale of Arsenic Regulation Act**: Barrell, ibid., pp.128–61. **1851 census**: Fox Hall: www.upminsterhistoryblog.wordpress.com. **Honorary MD**: University of St Andrews senate minutes, UYUY452/17. **Cases**: Lister: *Lloyd's Weekly Newspaper*, 7 November 1852; *Manchester Examiner & Times*, 10 November 1852. Note that Ann Skingsley might have been in fact Emma Skingsley. Kirwan: Boswell, John Knight, *Defence of William Bourke Kirwan*, Webb & Chapman, Dublin, 1853; www.pamlecky.com. Neligan: obituary in *The Lancet*, 1 August 1863. Pinckard: *Northampton Mercury*, 28 February 1852. Atlee case: *Leicester Chronicle*, 18 February 1854; *The Lancet*, 11 & 25 February 1854; *Chelmsford Chronicle*, 10 November 1854. March quarter 1854, there are deaths of five people called

Attlee registered in the Croydon District: Elizabeth, Harriet, John, and two Williams. Presumably this is the family that died. Burton case: *Sussex Advertiser & Surrey Gazette*, 23 May 1854. Image: obituary, *The Lancet*, 3 October 1903. Ogston: obituary, *BMJ*, 1 October 1887. **St James's Terrace:** *BMJ*, 10 October 1874, p.477. **Arsenical cake decoration:** *Associated Medical Journal*, vol. 1 (5), 4 February 1853, p.105. *Guy's Hospital Reports*, series 2, vol. vii; *Northampton Mercury*, 22 July 1848; Whorton, James C., *The Arsenic Century: How Victorian Britain was Poisoned at Home, Work and Play*, Oxford University Press, 2010, pp.151–68.

Chapter 11
Great Fire of Newcastle and Gateshead: Rewcastle, James, *A Record of the Great Fire of Newcastle and Gateshead*, George Routledge, London, 1855. Catherine O'Brien: *Huddersfield Chronicle*, 18 November 1854; detractors of Taylor's theory: *The Cumberland Pacquet* 31 October, *Newcastle Journal*, 11 November 1854. **Southgate:** Taylor, *On Poisons*, 1859, ibid., p.277. *Lloyd's Weekly Newspaper*, 24 December 1854; *Morning Chronicle*, 26 January 1855; *The Morning Advertiser*, 8 & 15 February 1855. Southgate's will: PCC PROB 11/2209/179. **Wooler:** marriages: Findmypast Northumberland and Durham Marriages and Westminster Marriages, *Sheffield Daily Telegraph*, 18 July 1855; *The Hertford Mercury*, 28 July 1855; *York Herald*, 4, 11 & 18 August, 1 September 1855; *The Times*, 10, 12, 17 (quoting *The Darlington Times*), 21, 22, 27 December 1855, 7 January 1856 (quoting the *Pharmaceutical Journal*). Charles Wilkins's background: Woolrych, Humphry William, *Lives of Eminent Serjeants-at-Law of the English Bar*, vol. 2, W.H. Allen, London, 1869, pp.850–88. Parke's letter: Heuston, R.F.V., 'The Wensleydale Peerage Case: A Further Comment', *The English Historical Review*, vol. 83, no. 329 (October 1968), p.781. Hydrochloric acid in bread: Graber, Kate, *Nebraska Pioneer Cookbook*, University of Nebraska Press, Lincoln & London, 1974, p.6. Von Tschui and arsenic-eaters: Burney, ibid., pp.66–8; Taylor, *On Poisons*, 1859, pp.94–8. **Legge v Edmonds:** *Hereford Times* 24 November, 1 December 1855; Tidy, Charles Meymott, *Legal Medicine*, vol. 3, William Wood, New York, 1884, p.49.

Chapter 12
Taylor: Taylor, Alfred Swaine, *On Poisoning by Strychnia, with Comments on the Medical Evidence Given at the Trial of William Palmer for the Murder of John Parsons Cook*, Longman, Brown, Green, Longmans & Roberts, London, 1856; chapter on strychnine in *On Poisons*, 1859, ibid., pp.672–705; on nux vomica and strychnine in *Medical Jurisprudence*, 1853 (US ed), pp.157–60; 'On poisoning by tartarized antimony', *GHR*, series 3 vol. 3, 1857, p369-481; 'On the detection of absorbed strychnia and other poisons,' ibid., pp.482–501; 'Analysis of the water of the Great Geyser, Iceland', ibid., series 3, vol. 2, pp.405–407. Taylor's correspondence with Gardner, and the Rugeley postmaster case: *The Times*, 12 January, 15 March 1856. Anon, *Illustrated Life and Career of William Palmer, of Rugeley*, Ward & Lock, London, 1856. The same company published *Illustrated and Unabridged Edition of The Times Report of the Trial of William Palmer*, Ward & Lock, London, 1856. The second book contained graphics from the *Illustrated Times*. **Robert Warrington:** *The Lancet*, 1842, vol. 38(989), pp.703–704. **Cook's background:** Chancery case: TNA C 13/2958/20; *Morning Chronicle*, 14 June

1836. His mother, Ann Stephens: Marriage in Eton reg district. Baptism of Edmund at St Giles-in-the-Fields, 11 July 1839, son of William Vernon and Ann Stephens of 5 Alfred Place, gentleman. Burial of Mrs Ann Stephens of 5 Alfred Place at Kensal Green, 1 June 1839, aged 36. Will of John Cook of Horton, Buckinghamshire, 1845 PCC PROB 11/2017/156. Will of John Parsons Cook of Worthing, Sussex, 1856 PCC PROB 11/2225/174. Biographical sketch: *The Cumberland Pacquet*, 17 June 1856, and *Illustrated Life and Career…*, ibid. Shrewsbury Autumn Races: *The Berkshire Chronicle*, 17 November 1855. **The inquests:** John Parsons Cook: The first report appears to have been an article in the *Staffordshire Advertiser*, 1 December 1855, 'Mysterious Death of a Sporting Gentleman'; inquest was opened and adjourned. Subsequent meetings were reported in the same newspaper on 15 & 22 December 1855. *The Times* first reported it on 18 December 1855, giving an account of the whole inquest. Entry on Rees in Wilks & Bettany, 1892, pp.251–61. Ann and Walter Palmer: *The Times*, 20 December 1855, 4, 12, 14, 16, 17, 24, 25 January 1856. **Palmer's finances, Mayhew's life assurance offices investigation:** Bates, Stephen, *The Poisoner: The Life and Crimes of Victorian England's Most Notorious Doctor*, Duckworth Overlook, London, 2015. William Palmer is the focus of Bates's book. **Palmer's family:** Staffordshire baptisms, marriages and burials at Findmypast. Joseph Palmer's death announcement, *Staffordshire Advertiser*, 15 October 1836. Anon, *Illustrated Life and Career …*, ibid. There are four baptisms for children at St Augustine's. The inquest into Ann Palmer's death mentions she had five confinements; one child was stillborn. **Cases:** Ashmall: *Berkshire Chronicle*, 12 April; *Staffordshire Sentinel*, 26 April 1856. Dove case: Browne, Lathom G. & Stewart, C.G., *Reports of Trials for Murder by Poisoning*, Stevens, London, 1883. pp.233–37. **Wilkie Collins:** Burney, ibid., p.188. Wilkie Collins worked at *The Leader* during the Palmer trial; Thomas Boyle in *Black Swine in the Sewers of Hampstead* claims he was their Rugeley correspondent. Discussion in the press post-trial is from Burney. John Sutherland argues in *Victorian Fiction: Writers, Publishers, Readers*, Palgrave Macmillan, Basingstoke, 2005, pp.28–54, that Collins was inspired by the Palmer trial to structure *The Woman in White* around circumstantial evidence. Sutherland based his knowledge of the Palmer trial on Rupert Graves's inaccurate depiction in *They Hanged My Saintly Billy*. **Taylor and the press:** *The Lancet*, 19 January p.78; 2 February pp.134–5; 29 March 1856, p.348. The *Illustrated Times*, 2 February 1856, pp.91–2; Smith's demands, *The Times*, 28 March 1856, *The Lancet*, 29 March 1856, p.348. **The trial:** *Old Bailey Proceedings Online*; Knott, George, H., revised by Watson, Eric R., *Trial of William Palmer*, William Hodge, London, 1952; *The Times*, 28 May 1856. Wilkins's escape: Woolrych, ibid. *The Annual Register 1856*, Rivington, London, 1857, pp.387–8. Herapath: Watson, K.D., 'Herapath, William (1796–1868)', *ODNB*. Merritt: Taylor, *On Poisons*, 1859, ibid. Taylor enjoyed the telling of Letheby's slip-up, but his account seems otherwise objective. **Post-trial strychnine discussions in *The Lancet*:** Wilkins's letters 30 May, 20 June; Rodgers & Girdwood's letters 13 June, 18 July; Letheby: 14 November 1857.

Chapter 13
Public health: Fleet Mills: *The Leeds Times*, 22 March 1856. Tinfoil: Taylor, *On Poisons*, 1859, p.450, *Gloucester Journal*, 21 July 1855. **Bacon:** *The Lancet*, 21 February, p.162; *The Cheltenham Chronicle*, 24 February; *The Grantham Journal*, 28 February, 1 & 28 August; *Isle of Wight Mercury*, 16 May; *The Times*, 29 July 1859. Taylor, *On Poisons*, 1859, p.157. **The Waterloo Bridge Mystery:** *The Times*, 10, 12, 13, 14, 16, 27 October, 5 November, 30 December 1857. Spurgeon, C.H., 'Murder Will Out', British Library, 1888.c.3(63). Smith, Sydney, ed., *Taylor's Principles and Practice of Medical Jurisprudence*, 10th ed., Churchill, London, 1948, pp.143–5.

Chapter 14
The Starkins case: *Dorset County Chronicle*, 19 November 1857; *Hertford Mercury*, 13 March 1858; *Herts Guardian*, 21 November 1857, 13 March 1858. Thomas Monk: *Fife Herald*, 10 December 1857; *Herts Guardian*, 19 December 1857; *Liverpool Mercury*, 19 February 1858. **Public health:** Hampshire: *Dorset County Chronicle*, 21 January 1858; Whitehaven's water, *Cumberland Pacquet*, 31 August 1858. **Arsenical pigments:** *Journal of the Society of Arts*, 1858, 9 January p.39, 27 August p.606, 17 September pp.636–9, 8 October pp.665–7, 31 December p.98; Taylor, Alfred Swaine, 'Ophthalmia as a result of the use of arsenical wallpapers', *Ophthalmic Hospital Reports & Journal of the London Ophthalmic Hospital*, no. 6, January 1859. Whorton, ibid., pp.202–28. **Surgical Instrument Committee:** *Journal of the Society of Arts*, 28 May 1858. **The Swiney Prize:** *British Medical Journal*, 12 February 1859. *On Poisons*: review in *The American Law Register*, vol. 7(9), July 1859, pp.573–5. **Smethurst:** Parry, ibid., Stephen, James Fitzjames, *A General View of the Criminal Law of England*, 2nd ed. Macmillan, London, 1890, pp.286–317. Summerscale, Kate, *Mrs Robinson's Disgrace: The Private Diary of a Victorian Lady*, Bloomsbury, London, 2013, p.61. Summerscale describes Smethurst in passing as 'the well-known hydropathist', but omits mention of his trial. Ballantine, ibid., pp.212–22. Burney, ibid., Taylor, Alfred Swaine, 'Facts and fallacies connected with the research for arsenic and antimony', *Guy's Hospital Reports*, series 3, vol. 6, 1860, pp.201–65. *Old Bailey Online*. Collins, Charles Allston, 'Small Beer Chronicles', *All The Year Round*, 3 January 1860, p.399. Dickens, Charles, 'The Modern Alchemist', *All The Year Round*, 27 December 1862, pp.380–4. Symons, Julian, *Bloody Murder: From the Detective Story to the Crime Novel: A History*, Penguin, Harmondsworth, 1974, pp.53–4. Collins, Wilkie, *Armadale*, Smith, Elder, London, 1872, p.520. Baker, William, *Wilkie Collins' Library: A Reconstruction*, Greenwood, Westport, 2002, p.155. Golan, Tal, *Laws of Men and Laws of Nature: The History of Scientific Expert Testimony in England and America*, Harvard University Press, London, 2004, p.102, 110–18. Genealogy compiled from records at Findmypast and Ancestry.co.uk.

Chapter 15
Hopley: Moore, Julian, 'Hopley, Thomas (1819–1876)', *ODNB*. *The Times*, 24 July 1860, 21 April 1960. Clive, Caroline, 'Beaten to Death', *The Constitutional Press*, vol. 3, June 1860. Hydrocephalus and learning difficulties: www.nhs.uk House of Lords Debate, 20 May 2004, vol. 661, cc.890–914. Taylor, Alfred Swaine, *The Manual of Medical Jurisprudence*, John Churchill, London, 1861, p.265, 291, 312.

Cases: Essex quackery: *Chelmsford Chronicle*, 8 June 1860. Wimbledon: *Morning Chronicle*, 18, 22, 24 February; *Bell's Weekly Messenger*, 1 March 1862. Constance Wilson AKA cases: *Old Bailey Online*; Williams, Montagu, *Leaves of Life*, vol. 1, Houghton, Mifflin, Boston & New York, 1890, pp.77–82. *The Times*, 16 May, 5 July, 29 September, 1, 21, 22 October 1862. Austin, W.S., 'Secret Poisoning', *Temple Bar*, vol. 6, November 1862, pp.579–84. Doige: *Royal Cornwall Gazette*, 15 August 1862. **Darwin letters:** DAR 58.1: 6 & 58.1: 14–15. Cambridge University Library.

Chapter 16
Chemistry: Brande, William Thomas & Taylor, Alfred Swaine, *Chemistry*, John W. Davies, London, 1863. Review: *BMJ*, 24 January 1863, p.92. **Body temperature:** Taylor, Alfred Swaine & Wilks, Samuel, 'On the cooling of the human body after death: inferences respecting the time of death. Observations of temperature made in 100 cases', *GHR*, series 3, issue 9, 1863, pp.180–211. Comments on: Knight, Bernard & Madea, Burkhard, 'Historical Review on Early Work on Estimating the Time Since Death', in Madea, Burkhard, ed., *Estimation of the Time Since Death*. Boca Raton: Taylor & Francis, 2016, pp.7–9. Information on Wilks from Cameron, ibid., pp.236–7. **Cases:** Bragg: *Reading Mercury*, 19 June 1864; Earls Colne: *Chelmsford Chronicle*, 9 November 1866; Clarke: *West Middlesex Advertiser & Family Journal*, 5 August 1865; Biggadyke: *Liverpool Daily Post*, 29 December 1868. Eldridge: *Cambridge Independent Press*, 1 August 1863; Taylor, A.S., *Principles and Practice of Medical Jurisprudence*, John Churchill, London, 1865, p.429; Aldgate pump: *London City Press*, 13 October 1866; River Lea: *Luton Times and Dunstable Herald*, 30 May 1868; Brande: *Proceedings of the Royal Society*, vol. 16, 1867–68, pp.ii–vi; magenta dye: Golan, ibid., pp.84–8; inquests: *Liverpool Daily Post*, 4 July 1862; Food Committee: *Journal of the Society of Arts*, 3 May 1867, pp.375–80; cattle plague: *Derbyshire Times & Chesterfield Herald*, 11 May 1867; *A Description of Spencer's Patent Magnetic System of Purifying Water: with authenticated testimonials and extracts showing practical results*, Magnetic Filter Company, London, 1869. Fanny Adams: *Sheldrake's Aldershot and Sandhurst Military Gazette*, 19 October; *Morpeth Herald*, 14 December, *Illustrated Police News*, 14, 28 December; *Western Gazette and Flying Post*, 27 December 1867; Flanders, Judith, *The Invention of Murder: How the Victorians Revelled in Death and Detection and Created Modern Crime*, Harper Press, London, 2011, pp.381–4. Waterer: *London Evening Standard*, 1 December 1863; *York Herald*, 2 January 1854; *West Surrey Times*, 30 January 1864. Suffolk: *North London News*, 5 March 1864. Briggs case: Colquhon, Kate, *Mr Briggs' Hat: A Sensational Account of Britain's First Railway Murder*, Abacus, London, 2012; Taylor, 1865, ibid., p.450. **Select Committee:** *BMJ*, 22 July 1865; *Birmingham Daily Gazette*, 20 October 1864; Royal Pharmaceutical Society, www.rpharms.com. Review of *Principles and Practice …*: *The Lancet*, 21 October 1865, pp.457–9. **Blood:** Wagner, E.J., *The Science of Sherlock Holmes: From Baskerville Hall to the Valley of Fear, the Real Forensics Behind the Great Detectives Greatest Cases*, Wiley, Hoboken, 2006, pp.169–71. Taylor, Alfred Swaine, 'On the detection of blood by guaiacum', *GHR*, series 3, vol. 19, 1874, pp.517–19; Taylor, Alfred Swaine, 'On Homicidal and Suicidal Wounds of the Throat', *GHR*, series 3, vol. 14, 1869, pp.112–44. **Photographs of Eminent Medical Men:** Robertson, William Tindal, ed., ibid. **Boulton and Park:** McKenna,

Neil, *Fanny & Stella: The Young Men Who Shocked Victorian England*, Faber & Faber, London, 2014; *The Lancet*, 25 June 1870, p.911; twopenny pamphlet: www.rcplondon. ac.uk (Royal College of Physicians).

Chapter 17
Retirement: *The Lancet*, 10 Sep 1870; *Chelmsford Chronicle*, 19 April 1872. **Work:** *Derbyshire Times & Chesterfield Herald*, 11 November 1871; Taylor, Alfred Swaine, *Metropolis Water Supply. Remarks on the water supplied by the West Middlesex Water Works Company*, 1872, London; *Journal of the Society of Arts*, 30 June 1871; *The BMJ*, 22 November 1873, p.606. Complaint: *The Lancet*, 4 April 1874. **Christison:** Dispute with women medical students: Roberts, Shirley, *Sophia Jex-Blake: A Woman Pioneer in Nineteenth-Century Medical Reform*, Routledge, London, 1993. Baronetcy: *The London Gazette*, 21 November 1871. **West Haddon Tragedy:** Taylor, Alfred Swaine, 'Death from disease or poison: Does the retention or maintenance of heat in a dead body furnish any indication of the cause of death?', *GHR*, series 3, 19, 1874, pp.468–87; *The BMJ*, 24 January 1874, p.113. **Tattoos:** Taylor, Alfred Swaine, 'Medico-legal observations on tattoo-marks as evidence of personal identity. Remarks on the Tichborne [*sic*] case', *GHR*, series 3, vol. 19, 1874, pp.441–65. **Regent's Park Explosion:** *The Times*, 3 October; *The Lancet*, 10 October 1874, pp.477–8. **Later work:** Quack medicine: *The Lancet*, 24 July 1875, pp.147–8; Vivisection: *BMJ*, 10 June 1876; *Arnott's Elements of Physics*: *BMJ*, 10 February 1877; 'The Penge Mystery': *The Gloucester Citizen*, 26 September, *Dundee Courier*, 13 October 1877, *The Guardian*, 15 April 2012; Coroners: *BMJ*, 19 January 1878. Meets Werge: White, Stephen, ibid., pp.234–5. **Death and obituaries:** Caroline Taylor's death: *Pall Mall Gazette*, 10 February 1876. Taylor's death: *The Times*, 29 May 1880; *BMJ*, 12 June 1880, pp.905–906; *The Lancet*, 5 June 1880, p.897; *Medical Times and Gazette*, 12 & 19 June 1880; Harveian Oration: *BMJ*, 31 July 1880, p.161; *Medico-Legal Journal*, vol. 3(1), 1885–86, p.140. **Legacy:** Stevenson, Thomas, ed., *The Principles and Practice of Medical Jurisprudence*, Henry Lea, Phildelphia, 1883, pp.v–vi. Doyle, L.R., 'Book Reviews', *Nebraska Law Review*, vol. 31(3), March 1952, pp.504–505. Mant, A. Keith, ed., *Taylor's Principles and Practice of Medical Jurisprudence*, Churchill Livingstone, Edinburgh, 1984. Pearsall, Ronald, *Conan Doyle: A Biographical Solution*, Weidenfeld & Nicolson, London, 1977, p.9, 56. Doyle, Sir Arthur Conan, 'A Study in Scarlet', *Sherlock Holmes: The Complete Stories*, Wordsworth, London, 2007, pp.15–17, 19, 574. Doyle, Arthur Conan, *The Stark Munro Letters*, Longmans, Green, London, 1895, pp.140–1. Freeman, R. Austin, *The Red Thumb Mark. With a new introduction by Otto Penzler*, Mysteriouspress.com/Open Road, New York, 2014 (ebook, unpaginated); Freeman, R. Austin, *The Famous Cases of Dr Thorndyke*, Hodder & Stoughton, London, 1941, p.23. Freeman on his readers quoted in Kayman, Martin A., 'The short story from Poe to Chesterton', in Priestman, Martin, ed., *The Cambridge Companion to Crime Fiction.* Cambridge University Press, 2003, p.47; Symons, Julian, ibid., pp.88–90. Sayers, Dorothy L., *Whose Body?*, Boni & Liveright, New York, 1923, and *Busman's Honeymoon*, Victor Gollancz, London, 1937, pp.253–4. Graves, Robert, *They Hanged My Saintly Billy*, Xanadu, London, 1989, p.233.

Index

Abortion, 8, 12, 15, 56, 91
Adams, Fanny, 186–7
Aikin, Arthur, 4, 16–17, 56, 97, 106, (note: 215)
Albemarle School, 1–2, (Plate 2)
All the Year Round, 164–5
Allen, William, 4
Amos, Andrew, 8–10
Anatomy Act 1832, 17
Antimony, 115–17, 141, 144
  Palmer, William and, 122–5, 127, 130, 134–8
  Reinsch test, and, 37, 129
  Smethurst, Thomas and, 155–62
Apothecaries, Worshipful Society of, 3, 4, 6, 7, 10, 40, 46, 155
Arsenic
  abortifacient, as a, 12, 56, 91
  dye, as a, 100–101, 152–4, 179
  Marsh test, 22–2, 32–5, 36–7
  murder trials, and, 45, 122
    R. v Bacon, 143–6
    R. v Biggadyke, 181
    R. v Chesham, 63–5, 68, 88–90
    R. v Dore and Spry, 73–6, 184
    R. v Geering, 80–1
    R. v Jennings, 48–50, 52–3
    R. v Leaver, 42–3
    R. v Lucas and Reeder, 85–6
    R. v May, 69–71, 73
    R. v Merritt, 139
    R. v Rhymes, 31–5
    R. v Russell, 12
    R. v Southgate, 76–7, 79–80
    R. v Wooler, 106–109, 110–12
  poisoning symptoms of, 24, 49, 70, 90, 107

public health, and, 184
Reinsch test, and 36–8, 108, 115, 120
  apparent unreliability at the Smethurst trial, 155–63
restriction on sale of, 83, 96, 153, 182
soil analysis and, 86
suicide and, 24–5, 101–102
Styrian peasants, and, 111
Arzone family, 153
Ashmall, Catherine, 132
Asiatic cholera, *see* Cholera, 'Asiatic'
Atlee family, 101

Bacon, Martha and Thomas Fuller, 143–6
Baker, Frederick, 186–7
Ballantine, William, 79–80, 85, 88, 168–9
  R. v Smethurst, and, 157, 160, 163
Bamford, William, 116–19, 123–4
Barry, Alexander, 16–17
Barry, James, 18, 192
Bate, George, 122, 136, (Plates 23, 24)
Beck, Theodric Romeyn, 4, 21, 41
Bell, Joseph, 200
Berkshire, county of, 31–5, 48–50, 52–3
Bigamy, 155–62
Biggadyke, Priscilla, 181
Bitter almonds, 51, 56
Blood analysis, 91–4, 95–6, 151, 174, 183, 185–7
Bowyer, Hannah, 69, 71–3
Brande, William Thomas, 131, 137–8, 158, 179, 188, (Plate 30)
Briggs, Thomas, 183
Brighton, 20, 21
British Medical Journal, 41, 162, 179, 190, 196, 197, 199
Brodie, Sir Benjamin, 133, 139, 161

Buckinghamshire, county of, 50–3
Bulwer-Lytton, Edward, 65–7
Burial clubs, 70–1, 77, 80–1, 129
Burke and Hare, 6–7, 17, 200

Calcraft, William, 52, 53, 73, 85, 86, 174, 178, 187, 196
Calomel, *see* mercurious chloride
Cambridgeshire, county of, 19, 85–6, 106
Campbell, John, 1st Baron Campbell, 96, 132, 139–40
Cancellor, Caroline, 19–21, 27, 30, 42, 190, 195
    family of, 19–20, 81–2, 170
        the Eastbourne Manslaughter, and, 166–70
Cancellor, John Henry, 166–9
Cancellor, Reverend John Henry, 166–8
Cancellor, Reginald Channell, 166–70
Carpenter, Jeremiah, 151
Central Criminal Court, *see* Old Bailey
Central Criminal Court Act (Palmer's Act) 1856, 127–8
Champneys, Henry Montague, 50–6
Chancery, Court of, 110, 113, 179
Chesham, Sarah, 63–5, 68, 71, 88–90, 95–6
Chester, 170
Chloroform, 45, 79, 189
Cholera and arsenic poisoning misdiagnosis, 31, 56, 70, 85, 88
Cholera, 'Asiatic', 14–15, 46, 82–5, 176
Cholera, 'English', 14, 88, 124, 144, 175–7
Christison, Robert, Dr later Sir, Baronet, 16, 40, 45, 65, 107, 110–12, 193, 200, 201
    arsenic analysis, and, 22–4, 33–4, 36–8, 163
    career of, 6–7, 57, 192
    influence on Sir Arthur Conan Doyle, 7, 200
    Palmer trial, and, 131, 138, 140–1
Churchill, John, 36, 41
Churchyards, overcrowding of, 29, 46, 82

Cockburn, Sir Alexander, 132, 138–9, 169–70
Colchicum, 176–7
Coleridge, Sir John Taylor, 33–4
Collins, Wilkie, 127, 165
Conan Doyle, Arthur, Sir, 7, 171, 199–201
Concealed sex, *see* LGBTQI issues
Consumption, *see* Tuberculosis
Cook, John Parsons, *see* Palmer, William
Cooper, Sir Astley, 4, 18
Cooper, Bransby, 42, 99
Cooper, John Thomas, 51, 53–4
Cornwall, county of, 58, 174
Coroners Act 1887, 196, 198
Corporal punishment, *see* Hopley, Thomas
County Constabularies Act 1839, 32
Cresy, Edward, 171
Crime fiction, 65, 127, 165, 191, 199–203
Crime scene, 186, 191
Cyanide (hydrocyanic cyanide, Prussic acid), 16, 50–6, 66, 106, 119, 123, 125–6

Daguerre, Louis, 20, 26, 28, 46
Darwin, Charles, 171, 195
Debus, Heinrich, 190
Dickens, Charles, 85, 94, 122, 164–5
Dissection, 3, 4, 17–18, 47, 148–9
Doige, John, 174, 181
Dore, Mrs, 73–6
Dove, William, 133, 136, 138
Doyle, Sir Arthur Conan, *see* Conan Doyle, Sir Arthur
Drory, Thomas, 91–6, (Plate 14)
Drowning, 98–100
Dublin, 98–100

Eastbourne Manslaughter, the, *see* Hopley, Thomas
Edinburgh, 6–7, 18, 45, 79, 82, 192, 199
    Christison, Robert, and, 6–7, 57, 65, 106–107, 192, 199–201, (note: 218), (Plate 4)
Edwards, Eliza, 17–18, 20, 189, 192

Eldridge, Alfred, 181
Enema, nutrient, 60, 107, 109, 111
English cholera, *see* Cholera, 'English'
Epsom salts, 104–105
Erichsen, John, 168–9
Essex, county of, 63–5, 69–71, 73, 76–8, 88–90, 91–8, 181, 190
Explosions, 16, 103–104, 194
Experiments, dangers of, 16–17, 22, 199
Experts, disagreements between, 54–5, 73–6, 140–2, 154–5, 185

Faraday, Michael, 23, 26, 28, 128, 136
Felix, Charles, 165
Felon literature, 65–8
Fibres analysis, 181, 184, 186
Finchley, Middlesex, 104–106
Food analysis, 38, 42–3, 101, 109, 143, 188, 190
Forensic entomology, 56
France, 2, 4–6, 35, 45, 55, 66, 77, 94, 132, 190
Freeman, R. Austin, 201–202
French, John George, 13–14, 45, 82

Gardner, Samuel, 180–1, 191
Gateshead, 103–104
Geering, Mary Ann, 80–1
Geoghegan, Dr Thomas Grace, 35, 98–9, 141, 163
Gorway, Charles M., 125, 128, 136–7
Graves, Robert, 203
Great Fire of Newcastle and Gateshead, the, 103–104
Grotta del Cane, 5, 16
Guaiacum, 95, 185
Guy, Thomas, 3
Guy, Dr William Augustus, 43–4, 110, 188
Guy's Hospital, London, 3–4, 10–11, 16–18, 24–5, 36, 79, 157, 164–5, 180–1
cholera epidemic and, 82–4
expert witnesses and, 12–13, 42, 52–4, 75, 87

Former pupils of, 28, 31, 54, 59, 62, 179, 189
Guy's Hospital Reports, 22–5, 31–2, 54–5, 62, 64, 100, 140–1, 162, 193–4

Hair analysis, 147, 181
Hampshire, county of, 94–5, 152, 186–7
Hartley, Louisa Susan, 87–8
Hassall, Arthur Hill, 101
Hemiplegia, paternity and, 109–10
Hemlock, 69, 71–3
Herapath, William, 53, 108, 154–5, 161–2, 171
Palmer trial, and, 131, 138–9
Herschel, Sir John, 27–9, 52
Hertfordshire, county of, 151
Highgate cemetery, 195, 197
Holmes, Sherlock, *see* Conan Doyle, Sir Arthur
Homeopathy, 157, 190
Hopley, Thomas, 166–70, 181
Household Words, 94
Huntingdon, 12
Hydrocyanic acid, *see* Cyanide
Hydrotherapy, 83, 156

Illustrated Police News, The, 187
Illustrated Times, 125, 128, 136–7, 142, 187
Image, William Edmund, 99
Inquests, flaws of, 46–7, 65, 84, 188, 193
Ireland, 98–100
Isle of Wight, 73–4

Jennings, Thomas, 48–50, 52–3
Jones, Charles Chadwick, 68, 71, 95
Jones, William Henry, 113–14, 116–20

Kelly, Sir Fitzroy, 53
Kent, county of 1, 3, 100, 181, 196
Kent, Constance, 191–2
King's College, London, 25, 43–4
Kirwan, William Bourke, 98–100
Knife wounds, 147–9, 151, 185–7, 191

Lafarge, Marie-Fortunée, 35, 66–7, 77, 90, 94
Landon, Letitia Elizabeth, 66
Laudanum, death from, 13, 184, (*see also* Opium)
Lancet, The, 7, 36, 55, 62, 101, 189, 191, 194
  opinion of Alfred Swaine Taylor, 13, 40–1, 43–4, 101, 162, 185
  medical jurisprudence teaching, and, 10, 15
  Palmer trial, and, 127, 130–1, 135–6, 140–2
Letheby, Henry, 45, 163–4, 179, 183, 188, (Plate 13)
  Tawell trial, and, 53, 55
  Palmer trial, and, 131, 138–9, 141–2
Lewis, Charles Carne, 63, 89, 92, 98
LGBTQI issues, 17–18, 103–104, 188–9, 192
Life assurances, 8, 13, 144
  the Palmer cases and, 121–2, 124–5, 128–9, 165
Lincolnshire, county of, 90, 144–5, 174, 181
Liverpool, 121–3, 165, 171, 184, 191
Locock, Sir Charles, 167–8
London Medical Gazette, 41, 43–7, 65–8, 79, 82–3, 84, 99
Lucas, Elias, 85–6

Mahaig, Joseph, 181–2
Manning, Frederick George and Maria, 83–5
Marsh, James, 22–5, 45
  see also Arsenic, Marsh test
Martin, Baron Samuel, 111–12
May, Mary, 69–71, 73, 76
Mayhew, Henry, 128–30, 136, 198
Medical education, 2–4, 17–18
Medical jurisprudence teaching, development of, 6–11
Medical Profession in All Countries, The, 191
Medical Times and Gazette, The, 197–8

Medical Witnesses Act 1836, 21
Medico-Legal Journal, 198
Mental illness, 144–5, 150, 187
Mercury, 37, 74, 117, 144, 155, 182
  mercurious chloride (calomel), 64–5, 76, 184
  mercury pills, 81, 151
Methold, Frederick John, 183, 196, 198
Miasma theory of disease, 9
  cholera, and the, 14–15, 83
Mills, Elizabeth, 117–20, (plate 26)
Monk, Thomas, 151
Morris, William, 153
Moxon, Walter, 193
Müller, Franz, 183

Neglect, 182, 195–6
Newcastle, 103–104
Nitro-benzole, 184
Norfolk, county of, 1, 29, 64
North, Mary, 59–62
Northfleet, Kent, 1, 29, 180, 196, (Plates 1, 9)
Nunneley, Thomas, 138, 174, 177, 188, (Plate 25)
Nux vomica, 94–5, 119, 140

Odling, George, 83–4
Odling, William, 83–4, 157–8, 162, 188, 190
Old Bailey, vi, 42, 73, 75–6, 88, 145, 158, 173–8
  the Palmer trial at, 128, 132–40
Opium (*see also* laudanum), 38–40, 41, 81, 83–5, 118, 184
Orfila, Joseph Bonaventura, 4, 6, 16, 24, 35, 52, 162
Oxalic acid, 6, 104–106

Palmer, Agnes, 122, 126
Palmer, Anne, 117, 120–5, 127, 129, 130, 141
  The Woman in White (novel) and, 165
Palmer, Walter, 121–6, 130, 136
Palmer, William, 113–42, 195

They Hanged My Saintly Billy, 203, (note 221)
Parke, Baron James, 72, 80, 112
Paternity, 109–10
Penge Mystery, the, 195–6
Pharmacy Act 1868, 184
Photogenic drawings, *see* Photography
Photographs of Eminent Medical Men, of All Countries, 187–8
Photography, 26–9, 179, 188, 196–7
Policing, 32, 70, 122, 123, 146–50, 151, 172–3, 183, 189
    detective techniques and, 42–3, 50–2, 92–3, 105, 143
Pollock, Frederick, 71, 73, 158–61
Potassium chlorate, 157–9, 163
Potter, Beatrix, 170
Prohibited degrees of marriage, 86, 97–8
Provincial Medical and Surgical Journal, see *British Medical Journal*
Prussic acid, *see* Cyanide
Public reaction to crime, 94, 161, 187

Quack medicines, 7, 45–6, 64, 83, 134, 157, 171
Quakers, *see* Society of Friends

R. v Bacon, 143–6
R. v Baker, 186–7
R. v Belaney, 55
R. v Biggadyke, 181
R. v Boulton and Parke, 188–9
R. v Bowyer, 69, 71–3
R. v Carpenter, 151
R. v Chesham, 63–5, 68, 88–90, 95–6
R. v Doige, 174, 181
R. v Dore and Spry, 73–6, 184
R. v Dove, 133, 138
R. v Drory, 91–6, (Plate 14)
R. v Eldridge, 181
R. v Gardner, 180–1, 191
R. v Geering, 80–1
R. v Hartley, 87–8
R. v Hopley, 166–70, 181
R. v Jennings, 48–50, 52–3

R. v Kirwan, 98–100
R. v Leaver, 42–3
R. v Lucas and Reeder, 85–6
R. v Mahaig, 181–2
R. v Manning and Manning, 83–5
R. v May, 69–71, 73
R. v Merritt, 139
R. v Müller, 183
R. v North, 59–62
R. v Palmer, 113–42, 185
R. v Rhymes, 31–5
R. v Russell, 12
R. v Smethurst, 155–65, 185
R. v Southgate, 76–7, 79–80
R. v Tawell, 50–6
R. v Wiggins, 185–6
R. v Wooler, 106–109, 110–12
R. v Wren, 94–5
Railways, 50–1, 68, 183
Reeder, Mary, 85–6
Regent's Park, London, 19–20, 100, 166, 183, 194
Rees, George Owen, 35, 45, 51, 93, 108
    Palmer cases and, 115–17, 122–8, 130–1, 133–4, 136–7
Reinsch, Hugo, 36–8
Rhymes, Hannah, 31–5
Rodgers, Julian, 74–6, 138, 141, 159, 163, 193
Rock Insurance Office and Kinnear, 13
Royal College of Physicians, the, 8, 10, 96, 154, 198
Royal Institution, The, 28, 137
Royal Society, The, 4, 16, 26, 44, 57, 58–9
Russell, Henry, 12, 15

Sale of Arsenic Regulation Act, 83, 96, 101
Sale of Poisons Bill, 152
Sayers, Dorothy L., 202–203
'Secret poisoning', 65–6, 77, 83, 128–9, 177
Sensation fiction, 127, 165, 178
Shee, Serjeant-at-Law, 132, 134–9

Skeletal remains, analysis of, 146–50
Smethurst, Thomas, 155–65, 185, 190, 195
Smith, John (solicitor of Palmer, William), 123–5, 127, 130–1, 136–7
Smith, John Gordon, 7–8
Smith, Robert Angus, 163–4
Snow, Dr John, 14, 45, 79, 83
Society of Friends, 17, 50, 52, 73
Sodomy, 17–18, 171, 188–9
Southgate, Hannah, 76–7, 79–80
Southgate, John, 104–106
Spry, Mrs, 73–6
St Andrew's, University of, 97, 156
St James's Infirmary, London, 13–15, 79
St James's Terrace, Regent's Park, London, 100
St Thomas's Hospital, London 3–4, 36
Staffordshire, 113–42
Stevenson, Thomas, 190, 192, 196, 198–9
Strychnine, 94–5, 133, 137, 182, 193
    tests for, 116, 119–20, 120, 130–1, 133–5, 138–42, 140–2, 164
Suffolk, county of, 54, 69, 71–3, 99, 182
Suicide, 12–14, 16, 24–5, 51, 54, 101–102, 104–106, 133, 141, 182
Sulphuric acid, death by, 13–14, 59–62, 87–8, 173–4
Surgical Instrument Committee, 154
Surrey, county of, 59–62, 84, 87–8, 155–6, 101
Sussex, county of, 80–1, 102
'Sweet Fanny Adams', 186–7
Swiney Prize, The, 154
Swinley, Mary Ann, see Taylor, Silas Badger
Syphilis, 114, 139

Talbot, William Henry Fox, 26, 28–9
Tattoos and identity, 193–4
Tawell, John, 50–6
Taylor, Alfred Swaine
    appearance of, 11, 125, 136–7, 191, 197, (Plates 11, 21, 24, 30–2)

birth of, 1
books by,
    Arnott's Elements of Physics (editor of), 195
    Chemistry, 179, 188
    Elements of Materia Medica and Therapeutics (editor of), 131
    Elements of Medical Jurisprudence, 21, 36, 40
    international readership of, 97, 191
    Manual of Medical Jurisprudence, A, 36, 40–2, 78, 97, 119, 154, 170, 171, 184–5
    On Poisons in Relation to Medical Jurisprudence and Medicine, 78, 86, 87, 105, 120, 152, 154–5, 157, 165, 171, 195
    On the Art of Photogenic Drawing, 28
    Principles and Practice of Medical Jurisprudence, 184–5, 191, 203
    Taylor's Principles and Practice of Medical Jurisprudence, 150, 198–9, 201, 202–203
    used in inquests and courtrooms, 53, 120, 123–4, 134–5
breadth of reading of, 45, 163
censuses, on the, 29–30, 96–7, 170, 190
Christian faith of, 59, 155, 197
courtroom procedure, opinion of, 34–5, 47, 52, 61–2, 67, 164, 199
daughter of, see Taylor, Edith Caroline
death and burial of, 197
education of, 1–6, 8
European travels, and, 4–6
fees of, 39, 84, 171–3, 196
geology and, 5, 16, 58–9, 140–1
home life of, 19–20, 27, 29, 81–2, 179–80, 194
laboratory at Guy's Hospital, 37–8, 164–5
lecturer, as a, 10–11, 17, 28, 179, 190
London Medical Gazette, and, 41, 43–7, 65–8, 82–4
obituaries of, 197–8

parents of, 1, 29
patents, negative opinion of, 26, 28–9, 46
photography, and, 26–9
popular press and, 94, 128–30, 136–7, 177–8, 187
posthumous legacy of, 197–203
public health, and, 31, 45–6, 82–3, 106, 109, 143, 152–4, 188, 190, 194
retirement, 190, 196
son of, 20
temperature, research and interest in, 48, 58–9, 180–1
wife of, *see* Cancellor, Caroline
will of, 196–7
writing process of, 41–2, 197
Taylor, Catherine, *see* Wilson, Catherine
Taylor, Edith Caroline, 42, 81, 97, 183, 196, 198
Taylor, Richard Alfred, 20
Taylor, Silas Badger, 1, 29–30, 97, 180, 190
    Swinley, Mary Ann (his wife), and 20, 21, 29–30
Telegraph messages, 50, 68, 84
Temperature, 48, 58–9, 93, 180–1, 193
Temple Bar, 177–8
Tetanic spasms, 107, 118–19, 133
The Times, 53, 64, 67, 77, 112, 146, 148, 169, 177, 197

William Palmer cases, and, 125, 127, 130, 139–41
Thomson, Anthony Todd, 8–11
Tindal, Nicholas Conyngham, 183
Titchborne Claimant, The, 193–4
Tuberculosis, 88–90, 196

Vitriol, oil of, *see* Sulphuric acid
Vivisection, 133, 135, 138–9, 195

Wakley, Thomas, 7, 36, 45–6, 105–106
Ward, William Webb, 116, 120, 123, 125–6, 132, 137, (Plates 22, 24)
Waterloo Bridge Mystery, the, 146–50
Werge, John, 196–7
West Bromwich, 56
West Haddon Tragedy, the, 193
Wiggins, John, 185–6
Wilks, Samuel, 180–1, 185, 193
Wilkins, Reverend, 70
Wilkins, Serjeant-at-Law, 111, 127–8, 132
Williams, Montagu, 173–4, 177
Wilson, Catherine, 173–8
Wilson, Constance, *see* Wilson, Catherine
Wooler, Joseph Snaith, 106–109, 110–12
Wren, William, 94–5